Teak Farming
(Tectona Grandis)

Teak Farming
(Tectona Grandis)

DR. K.K. JHA, IFS
Forest Department
Uttar Pradesh

CBS Publishers & Distributors Pvt. Ltd.

New Delhi • Bengaluru • Chennai • Kochi • Mumbai • Pune
Hyderabad • Kolkata • Nagpur • Patna • Vijayawada

Teak Farming (*Tectona grandis*)

ISBN: 978-81-239-2686-5

First CBS Reprint: 2015

Published by:
Satish Kumar Jain for CBS Publishers & Distributors Pvt. Ltd.,
4819/XI Prahlad Street, 24 Ansari Road, Daryaganj, New Delhi - 110002
delhi@cbspd.com, cbspubs@airtelmail.in • www.cbspd.com
Ph.: 23289259, 23266861, 23266867 • Fax: 011-23243014

Corporate Office: 204 FIE, Industrial Area, Patparganj, Delhi - 110 092
Ph: 49344934 • Fax: 011-49344935
E-mail: publishing@cbspd.com • publicity@cbspd.com

Branches:
• *Bengaluru:* 2975, 17th Cross, K.R. Road, Bansankari 2nd Stage, Bengaluru - 70
  Ph: +91-80-26771678/79 • Fax: +91-80-26771680
  E-mail: cbsbng@gmail.com, bangalore@cbspd.com
• *Chennai:* No. 7, Subbaraya Street, Shenoy Nagar, Chennai - 600030
  Ph: +91-44-26681266, 26680620 • Fax: +91-44-42032115
  E-mail: chennai@cbspd.com
• *Kochi:* 36/14, Kalluvilakam, Lissie Hospital Road, Kochi - 682018
  Ph: +91-484-4059061-65 • Fax: +91-484-4059065
  E-mail: cochin@cbspd.com
• *Mumbai:* 83-C, Dr. E. Moses Road, Worli, Mumbai - 400018
  Ph: +91-9833017933, 022-24902340/41 • E-mail: mumbai@cbspd.com
• *Pune:* Bhuruk Prestige, Sr. No. 52/12/2+1+3/2,
  Narhe, Haveli (Near Katraj-Dehu Road Bypass), Pune - 411041
  Ph: +91-20-64704058/59, 32342277 • E-mail: pune@cbspd.com

Representatives:

• Hyderabad: 0-9885175004	• Kolkata: 0-9831437309, 0-9051152362
• Nagpur: 0-9021734563	• Patna: 0-9334159340
• Vijayawada: 0-9000660880	

Printed at:
India Binding House, Noida (UP)

Preface

Teak is a native tree of Indian subcontinent. In past one hundred and fifty years it has spread to several tropical countries situated on both the sides of equator. It is extremely popular throughout the world for almost all kinds of timber uses. Mature tree produces quality timber between sixty to eighty years of age due to its moderately slow growing nature.

In recent years a number of private enterprises have come up for teak growing in many tropical countries including India. Guiding forces behind this have perhaps been initial fast growth and successful agroforesry practices resulting in early returns. Mushrooming private companies are offering shares to public and promising very high returns in twenty to thirty years. This requires detailed study on technical aspects and financial feasibilities mainly dependent thereon.

Irrespective of the growth performance in quick time teak has been a successful farm forestry crop. Heavy demand, big gap in supply and high price rise produce sufficient opportunity to go for it in a big way specially on private farms. This book is an effort to bring forth the growth and yield potential, to present the crop improvement, plantation and tending techniques and to assess the nutrient cycling in young teak plantations.

I am thankful to my wife, Neeru, and daughters, Sonal and Radhika, for their self-denial.

Well wishers, specially Mr Rai are also acknowledged for their support.

Dr K. K. Jha
Forest Deparment, Uttar Pradesh.
Lucknow; August 15, 1998

General Information on Teak

Scientific Name and Family:
Tectona grandis Linn., Verbenaceae

Vernacular Names :
Echingjagu, Segun, Saga, Sagch, Sagun, Sagaun, Jadi, Sagwani, Tega, Tyagadamara, Tekka, Thekku, Sagwan, Singuru, Tekku, Tekkumaram, Advitikku, Sag, Pedatekku.

Distribution :
Natural : India, Myanmar, Thailand, Laos.
Manmade : Both the sides of equator between 28°N and 18° S in all the continents except Europe.

Uses:
Lumber for ship building, Furniture, General carpentry.

Intercrops :
Suitable : Hill rice, Chilies, Tapioca, Ginger, Horse gram and Ragi.
Less suitable : Cotton, Maize, Sesame.
Non suitable : Irrigated rice, Plantain, Jute, Yams, Cocoyams, Eggplant, Creeping vegetables.

Mixed plantation species:

Bombax ceiba, Prosopis, Dalbergia sissoo, Albizzia lebbeck, Bamboos etc.

Damaging agency:

- Fire, Mistletoe, Theft.
- Insects : *Hybloea purea, Hapalia macharalis*

Yield record :

Forest grown

At 20 years - 0 to 68.5 m³/ha
At 50 years - 6.2 to 351.2 m³/ha

Plantation grown

At 20 years - 59 to 360 m³/ha
At 50 years - 200 to 480 m³/ha*

---

* In Costa Rican condition

Contents

Contents

Chapter 1

Introduction

Chapter 1

Introduction

eak (*Tectona grandis* Linn. f.) is a large deciduous tree. It is native to Southeast Asia. Its size generally varies from locality to locality depending upon the teak favouring quality of the site. The bole is tall, straight, fairly clean, cylindrical and with age moderately fluted and buttressed. Quadrangular and channelled branchlets support very large leaves which are broadly elliptical and obovate with entire margin. White flowers borne in dichotomous cymes of erect terminal panicles produce very hard fruits enveloped in bladder like structure.

This species is pronounced light demander, intolerant of shades and requires complete overhead light. It is very sensitive to mutual root competition and to suppression by weeds. It is tender to frost and severe drought. The tree is wind firm and fire resistant. It has remarkable ability to regenerate by coppicing and pollarding. The leaves are nonpalatable to grazers and browsers but attacked by insects. Roots are liked by rats and pigs.

The teak wood is literally a paragon among timbers. It is regarded as a very suitable species for the rapid production of large volumes of timber, poles and fuelwood. It is widely used as lumber for ship building, furniture and general carpentry (Akindele, 1989; Weaver, 1993). Many authors have described its uses in detail (Purkayastha, 1985; Weaver, 1993; Kadambi, 1993; Negi, 1996 and Tewari, 1992).

The reputation of teak is due to its matchless combination of qualities - termite, fungal and weather resistance, lightness with strength, attractiveness, workability and seasoning capacity without splitting, cracking warping or materially altering shape. The sapwood is white to pale yellowish brown, and heartwood pale brown to dark golden yellow, turning with age to brown, dark brown (Kadambi, 1993).

Teak grows naturally from approximate latitude of 23° to 10° N in Southeast Asia, an area of approximately 32 million hectares that encompass most of peninsular India, much of Myanmar, and parts of Laos and Thailand (Weaver, 1993). In India alone teak forests are distributed over 8.9 million hectares of land (Seth and Kaul, 1978).

The great adaptation capability, luring economics and ecological requirements of the area, convinced the foresters to take up teak plantation in teak as well as non teak zone. Several countries laid' emphasis on this exotic species in extensive as well as intensive forest management. Today teak is naturalized in many countries of Asia, Africa, Australia and

Latin America extending between 28° N to 18° S. Many million hectares of forest and nonforest lands are under teak cultivation. Table 1.1 indicates an over all picture of the popularity of teak in its favourable zones on global scale.

Today total area under teak plantations in the world is estimated at 3 million ha (Centeno 1997). India has over 0.5 million ha of teak plantations and massive ongoing programme to plant up about 50000 hectare annually (Khullar, 1995). In Uttar Pradesh Tarai region being the most favourable area for teak growth, this species was considered as one of the replacements of low productive Moist Deciduous Sal Forest as the part of intensive management. In a span of only thirty five years (1956 - 1990) more than 30000 ha area was planted with this species, the bulk of which is in Haldwani area.

Table 1.1 : **Distribution of natural and manmade Teak in the world (adopted from Hedegart, 1975 and Keogh, 1979 in Tewari, 1992).**

Region	Country
Southeast Asia	: India, Myanmar, Thailand, Laos, Indonesia, Cambodia, Nepal, Sri Lanka, Bangladesh, Pakistan, Vietnam Malaysia, Philippines, Papua New Guinea, Japan, Taiwan.
Pacific	: Australia, Solomon Island, Fiji Island, US Pacific Island.

East Africa	:	Tanzania, Sudan, Somalia, Zimbabwe, Uganda, Kenya, Malawi.
West Africa	:	Senegal, Guinea, Ivory Coast, Ghana, Togo, Dahomey, Nigeria.
South Africa	:	South Africa.
The Caribbean	:	Cuba, Puerto Rico, Panama, Honduras and others, Trinidad / Tobago, Jamaica, Nicaragua.
South America	:	Surinam, Brazil, Argentina, Colombia, Venezuela.
Central America	:	Belize, Costa Rica, El Salvador.

However, in the recent years the environmentalists have started raising fingers against this species apprehending that monoculture of teak tends to degrade the environment. It does not support wild life and ground vegetation and, therefore, biological diversity is also affected. It is feared that soil nutrient status and moisture regime are also affected adversely .

Irrespective of the controversy Teak is the most desired timber. Growth and quality of this species in plantation largely depends on soil characters, climate and management practices (Centeno, 1997) which includes latest impetus on intercropping. Therefore, this book is intended to discuss the prospective of teak growing as farm forestry or agroforestry emphasizing mainly on nutrient budgeting, potential yield, eliciting effect on production through various measures to be taken during pre and post planting period.

Chapter 2

Farm Forestry

Chapter 2

Farm Forestry

Teak (*Tectona grandis*) is the most sought after tree species for its fast production of large volume of timber and poles used in ship building, furniture and carpentry. Its popularity and wide range of adaptability has resulted in introduction of this species in various parts of the tropical world for extensive as well as intensive management of forestry. In India also it is being grown outside its natural zone at many places. Several thousand hectares of Tarai forests have been converted into teak plantations in Uttar Pradesh (UP) in past four decades. It has shown excellent performance in this region. It could prove to be a most promising species of farm forestry as well as agroforestry in the adjoining states having suitable climate and soil. Different related aspects of farm forestry are discussed below.

(a) Growth and yield

Commercially three distinct types of teak have been

recognized - forest grown, forest plantation grown and orchard teak (Reddy, 1995). It has also been reported that orchard grown teak attains six times more growth than forest grown teak and two times more than forest plantation grown teak. Therefore, it is assumed that yield of teak in Tarai orchard or similar ecological conditions should be about double the growth as recorded by Jha and Singh (1997) in plantation grown teak (54.07 x 2, 126.09 and 210.15 x 2 m³/ha at the age of 5, 18 and 30 years, respectively) or six times the growth reported by Kadambi (1993) in forest grown teak (68.5 x 6 m³/ha at the age of 20 years) of site quality one.

In teak orchard higher growth rate is achieved due to better aeration and irrigation during non rainy season, timely fertilizer application and protection from litter burning and fire damage (Reddy, 1995). Since trees once established do not respond significantly to irrigation and fertilizer application (Rawat, 1995) these operations should be carried out only for initial 3 to 5 years. The best way to ensure this is to go in for intercropping with agricultural components in the beginning and shade loving cash crops later. This will ensure irrigation, hoeing and fertilizer application in earlier years and in turn benefit the tree crop. This has been observed in Teak and Poplar plantation (Dagar *et al*, 1995; Jha and Gupta, 1991; Lahiri, 1989). In Tarai area of Nainital district wider rows of teak planting is in practice to utilize the interstrips for growing cereals and pulses. This has shown very promising results.

At very young age, especially at nursery stage, fertilizer, application helps the seedlings to fight deficiency problem in addition to growth improvement. Fernando (1966) has reported cure of deficiency symptom in teak seedlings suffering from nutrient deficiency. They responded to N, P, K fertilizers. N in organic form showed better results than in inorganic form. The symptoms were shortened internodes, small leaves, stunted growth and yellowing of seedlings.

(b) Wood quality

Additional water availability induces higher growth and wider growth rings which results in reduced timber strength (Kulkarni, 1994). Generally it is apprehended that fast growing and young aged teak is poorer in strength as compared to the naturally growing old aged forest teak. Tests at Forest Research Institute, Dehradun suggest that weight per unit volume and not the growth rate (faster or slower, per unit time) matter in the strength of timber. It is also suggested that plantation timber having wider growth rings is in no way inferior to naturally grown timber with thinner rings (Reddy,1995). In Nilambur teak Bhat (1995) has concluded that there is no large difference in wood density between younger and mature teak.

(c) Private venture

Foreseeing the prospects of teak farming a number of entrepreneurs have ventured into teak plantations on farmland in past few years. They are promising very high financial returns in twenty years. But many apprehensions have crept in against the growth projections and unfortunately these have not yet been over ruled due to the reasons that green gold venture is in its infancy and things are based on theoretical assumptions. Scanty experimental / research results also are not in their favour.

Mushrooming advertisements of green gold gimmick, especially teak plantation have attracted many professionals. They assessed the growth potential of this species but none of them (Kulkarni, 1994; Chaturvedi, 1995; Gogate *et al*, 1995; Kinhal, 1995; Parameswarappa, 1995 and Prasad, 1995) are convinced of highly projected returns of teak on various accounts. Growth of any species can not be improved beyond a limit because phenotypic expression depends not only on environmental factors but also to a large extent on genetic make up of the plant species and that the trees respond to external factors in initial few years till they are established.

The offers of green equities are heavily lopsided. This finding is based on evaluation of financial as well as technical aspects of some projects (Kinhal, 1995). There is a warning also for the investors about the pitfalls of putting money into teak plantations (Kulkarni, 1994). Comparative growth data have been presented in Table 4.1 to show a huge gap between recorded growth in plantations (a, b, c & d) and projected growth by the companies (w, x, y & z).

Table 4.1 : Comparative growth of teak at the age of 20 years.

	Volume (m³/ha)	References
	Recorded	
a	59.47	Kulkarni (1994)
b	68.50	Kadambi (1993)
c	140.00	Jha and Singh (1997)
d	240 - 360*	Centeno (1996)
	Projected	
w	800 - 900*	Centeno (1996)
x	972.00	Chaturvedi (1995)
y	1192.30	Kulkarni (1994)
z	2016.30	Kulkarni (1994)

* In Costa Rica

Status of Teak shares in Netherlands

A case study of a teak plantation company might benefit the readers as it is an eye opener. A Dutch owned company (FF) established approximately 3000 ha teak plantation in Costa Rica in 1989 and sold its economic ownership of the timber to another company (OH). The latter projected 800 - 960 cubic meter / ha commercial timber under 20 year cycles. It promised investors rate of return from 15 - 25 percent. On critical evaluation these projections were found to be questionable. The rate of return were based on unrealistic projections of timber yields and exaggerated expectations on the prices at which the timber could be sold. The company has been involved in legal complications and public scandals due partly to exaggerated claims on expected rate of return. Major flaw is extrapolated yield of timber after twenty years. The company is facing serious difficulties in substantiating its claims (Centeno, 1996).

Conclusion

Nonetheless, irrespective of the results of green equities, this much is assured that at the age of thirty years 210.15 m^3 timber/ha or 0.65 m^3 timber/tree can be obtained from plantations in Tarai or similar conditions (Jha and Singh, 1997). Even this amount of production is quite profitable seeing the steep increase in price of high grade sawn timber (from Rs. 700/- per cubic metre in 1960 to Rs. 50,000/- in 1995; Refer: Parameswarappa, 1995). Moreover, creating better environment like as the one in orchard and by intercropping it is possible to enhance teak growth significantly without affecting timber quality. Further this may result in reduction of rotation age from thirty to twenty five years assuring thereby the earlier returns. In addition agricultural and cash crop production will be the extra benefit.

Ecological considerations also go in favour of teak since it is neither more nutrient demanding nor less nutrient conserving. Teak produces considerably higher biomass at intermediate rate of annual productivity. Production potential of this species is higher than Tarai Sal and Central Himalayan forests.

Foregoing details give enough indications that teak farming in future will definitely improve social status of the people as well as ecological conditions of the area.

Chapter 3

Agroforestry

Chapter 3

Agroforestry

Forestry species are grown in combination with many agriculture crops at least in initial few years of establishment. Wide interspace having no shade is utilized as strips of agriculture field. This intercropping practice in early years of the plantations may have several beneficial effects e.g. (i) increased utilization of sites, (ii) benefits to the local community and (iii) maintenance of the area free of the weeds (Keogh 1987). The following two methods are common in many teak growing areas:

(a) *Taungya* System :

Taungya, a vernacular word from Burma, essentially means raising of a forest crop in conjunction with agriculture crops. This is practiced in several teak growing countries of the world. In India it is accepted system for establishing teak in Uttar Pradesh, West Bengal, Kerala, Assam and Tamil

Nadu. Occasionally it is adopted in Orrisa, Madhya Pradesh, Maharashtra and Mysore (Kadambi, 1993). The system varies slightly from country to country, but usually the cultivation of the field and forest crops is a joint effort of the land owner, generally the government, and a private farmer.

This system, introduced by the colonial government as early as 1875, was found a good proposition economically since it saved payment of wages and it could ensure a far better plantation survival rate. The system also helped them avoid hostility of local dwellers because of their utter dislike towards teak then known to them as highly useless species (Sinha, 1996).

Cultivation Method :

After clear felling of the forest, ground preparation is done first of all. This involves clearing of the site and burning of slash. Efficient burning is important for success of the crop. Teak is mostly introduced by stump planting or in some states by seed sowing in rows approximately 5 m apart. Strips between the teak rows, interspace, is used for growing field crops. Most common crops are rice, cotton, maize, sesame, wheat, ragi, pigeon pea, cassava, tomato and various other vegetables. Experiments have shown that suitable crops with teak are hill rice, chillies, tapioca, ginger, horse gram and ragi while cotton, maize and sesame are less suitable. Usually, nonsuitable crops are irrigated rice, plantain, jute, creeping vegetables, yams, cocoyams and eggplant. These field crops are grown 30 to 45 cm far off the teak row in first, second and in some states third year also. Weeding and tending of teak are done by cultivator during growth of field crop. Removal of forks and double leaders are also practiced.

In Uttar Pradesh, Mahrashtra, Orissa and parts of Kamataka and Kerala it is considered that *taungya* raised teak is better than those without *taungya*. However, it is reverse in West Bengal, Assam and parts of Tamil Nadu and Kerala. *Taungya* plantations are cheaper in the area where labour power is in abundance. Due to intensive and repeated soil working needed for agriculture teak plantation have minimum undergrowth.

(b) Group planting system :

This is an extension of the *taungya* system. Instead of the trees spaced equally throughout the plantation they are planted in groups. 250 groups of 8 trees per group or 2000 plants per hectare are planted in 6.3 meters square spacing. Within groups plants be arranged at the corners of two squares, the inner having a side of 1 m and the outer of 2 m. The advantage of this system over *taungya* is that the land may be utilized for longer period and perennials like banana may be interplanted (Keogh, 1987).

Taungya in West Bengal

Lahiri (1989) has reported teak regeneration by *taungya* method and impact of intercrops on teak growth. Performance of agroforestry crops has also been discussed. Some of his findings are as follows:

(i) The growth of forest crop improved in the intercropped area than non-intercropped area.

(ii)   The yield of intercrop was more under wider spacing (4m x 4m as compared to 3m x 3m). Growth of widely spaced teak also remained unaffected in long run, precisely at 10 years interval.

(iv)   Leguminous crops like soyabean, tentil, gram etc. did not perform satisfactorily owing to the acidic nature of soil.

(v)   *Citronella*, an aromatic herb, performed very well upto six years. Turmeric could be grown upto eight years, though yield decreased in later years. This decrease in yield was due to lowering of nutrient status of soil supporting regular cultivation.

Agroforestry in Haryana

Outside teak and *taungya* practicing region an experiment was conducted to identify suitable crops for growing in interspaces of six year old Teak plantation on reclaimed salt affected soil. It was found that barseem and gram could be grown successfully with Teak. Other crops like rice, wheat, pearl millet, mustard, sorghum and lentil were not found suitable companions. However, teak was benefited by these crops due to additional water availability through irrigation (Dagar *et al,* 1995).

PLATE 1 Agroforestry Practices

Wheat with Teak

Lentil with Teak

Chapter 4

Climate and Soil

Chapter 4

Climate and Soil

*T*eak is found in fairly dry areas experiencing excessive heat and prolonged drought in hot season. It dominates in drier part of its range but tends to get replaced in very moist tract. However, teak does not thrive under too hot or too moist conditions. It is well accepted fact that rainfall has a profound impact on teak distribution. As a matter of fact heavy moisture increases the growth. In lower rainfall areas teak growth is poor.

This species grows best in warm, moderately moist conditions. Optimum range of rainfall is 1500 - 2000 mm / year, but it endures rainfall as low as 500 mm/yr and as high as 5100 mm/yr. There is wide variation in adaptable temperature. This ranges from 2° C to 48°C. It also extends into the region that experience slight frosts (Weaver, 1993).

In Indian peninsula it grows on granite, gneiss, schists and other metamorphic rocks in Karnataka and Kerala. It is also found on limestones of Madhya Pradesh and Karnataka. It thrives well on Vindhyan sandstones but becomes stunted on quartzite or other metamorphosed sandstones which weather slowly.

As above *Tecton* (the carpenter's timber) grows on a variety of geological formations and soils, but its best growth is on deep, porous, fertile, well drained, alluvial soils with neutral or acidic pH. Teak tolerates wide extremes in soils as long as they are adequately drained. The most important limiting soil factors are shallowness, hard pans, water logged conditions, compaction, or heavy clays with low contents of Calcium or Magnesium. It has also been shown to be sensitive to phosphate deficiency (Weaver, 1993). The lateritic or very dry sandy soils do not support good growth to teak.

As a thumb rule teak can be grown in almost all non-refractory areas which can support at least low productive agriculture and have normal rainy season or irrigation facilities and proper drainage. However, its best growth is reported in warm tropical climate with 3 - 5 dry months and 15 - 20 cm annual rainfall and 22 - 27 °C temperature.

One of the best teak growing regions of the country Haldwani possesses typic hapludoll, fine loamy, mixed hyperthermic or typic euthrocrept, fine clayey, mixed hyperthermic soil. Clay and silt percentages are 21 - 50 and 22 - 10, respectively. Soil pH and EC in different aged teak stands range from 6.31 to 7.64 and from 0.03 to 0.14, respectively. Values of chemical attributes like Organic Carbon (OC), Organic Matter (OM), Nitrogen (N), Potassium (K) and

Phosphorus (P) are between 0.69 - 1.47, 1.19 - 2.55, 0.06 - 0.13, 0.0006 - 0.005 and 0.1 - 0.14, respectively. C/N ratio which is an indicator of fertility varied between 9.9/1 and 14.5/1 (Jha, 1995).

Nature and properties of some introduced teak growing soils of North - West Bengal were studied by Gangopadhyay (1987) in order to evaluate the characteristics of soil supporting this species. Structure is granular to fine subangular blocky, texture is loam to sandy clay loam and nature is acidic (pH from 4.5 to 5.5). Percentage ranges of OC, Total N, Total P_2O and Total K_2O are 2.5 - 3.5, 0.2 - 0.3, 0.2 - 0.4 and 1 - 3, respectively. Since teak is growing well in the area, it is assumed that the characteristics and nature of the soils must be favourable for the growth and development of the species.

Chapter 5

Nutreient Budgeting

Chapter 5

Nutrient Budgeting

Nutrients management has been well recognized since early times. This has become highly relevant with the advent of commercial forestry, where there is always a thrust to increase the production and biomass removal besides maintaining the site fertility. Therefore, there is a need to understand the nutrient distribution in various plant parts and also the nutrient cycling, especially in short rotation plantation ecosystem (Tandon *et al,* 1996).The cyclic movements of nutrients in ecosystems can be assigned to geochemical, biogeochemical and biochemical cycles. The latter two are much more relevant in the case of nutrient budgeting in teak.

(a) Biogeochemical Cycle :
(Resource Allocation Through Nutrient Cycling)

Plants accumulate resources and allocate them to growth, maintenance and reproduction through nutrient cycling. The process includes nutrient uptake, their retention in non photosynthetic biomass and return through litter fall. Retention or nutrient accumulation at a particular time is standing state. Apart from accumulation rate this depends on nutrient concentration.

(i) Nutrient concentration :

Concentration pattern of different nutrients in different plant parts are found to be different . Order of total nitrogen are : leaf > seed > lateral roots > fine root > tap root > bark > twig > bole = branch. Total phosphorus decreases in following order : lateral roots > seed > tap root > leaf > bole > bark > twig = fine root> branch. The Potassium is in following order : leaf > root > Branch > bark > bole. N / P ratio is maximum in leaf and minimum in bole.

(ii) Standing state :

Nutrients are collected generally from soil through uptake and through internal mobilization and they are used for structural building of the trees. Total accumulation in different plant parts is known as standing state of nutrient at particular time. Depending upon concentration of nutrients and amount of biomass production total storage of nutrients varies from one area to another. In Haldwani Tarai of U.P. range of above ground accumulation of nitrogen and phosphorus is 12.5 - 462.4 and 2.8 - 167.9 kg/ha, respectively (Jha, 1995). The age range is between one and thirty years. In another finding N, P, K, Ca and Mg content at the age of thirty in same area are 422.1, 40.6, 455.6, 992.4 and 287.6 Kg/ha, respectively (Negi *et al*, 1995). Density in this case is little higher than the previous stands. In Coimbatore forest nutrient storage at the age of 20 years in the form of N, P and K are 1370, 82 and 725 kg/ha, respectively (George and Verghese, 1992).

(iii) Nutrient uptake :

To meet the physiological requirements trees take up nutrients generally from soil. Gross uptake range of N and P between one to thirty years is 27.55 - 118.8 and 3.12 - 18.15 kg/ha /year, respectively. After correction for retranslocated amount the same is 18.83 - 79.73 and 3.01-16.06 kg/ha/year, respectively. George and Verghese (1992) have reported uptake of N, P, K at the age of 20 years 264, 17 and 132 Kg/ha/year, respectively, in Coimbatore forest.

(iv) Nutrient return :

Nutrients are returned on the forest floor through litter in the form of leaves, twigs and fruits. The quantity of return is maximum in leaves followed by twigs and fruits. The range of total return is 25.8 - 91.3 kg/ha/year for N and 2.74 - 10.09 kg/ha/year for P at the age of 30 years. Higher return has been reported by George and Verghese (1992) at the age of 20 years in Coimbatore area. N, P and K return in this case are 201, 15 and 84 Kg/ha/year.

Availability of nutrients on the forest floor or in the nutrient pool depends on its release through litter decomposition. Decomposition is a substrate dependent property which itself depends on flora and fauna. During the progress of decay colour alteration in litter, variation in nutritional status and change in fungal flora is found to be very distinct. Decay rates indicate that approximately 54 %

teak litter decomposes in first six months, starting from July. As regards the nutrient concentration nitrogen and phosphorus increase with the progress of litter degradation. Approximately one year time is sufficient for mineralization of nitrogen and phosphorus. In Nigerian climate this period is six months (Egunjobi, 1974).

(b) Biochemical Cycle :
(Mobilization Through Internal Cycling)

Metabolically active leaves help to conserve nutrients within stands through redistribution and prevent its loss through leaf fall. These leaves continue to drive nutrients till maturity and thus, control internal cycling of nutrients or retranslocation. In teak stands right from the day of sprouting, generally in April, leaves keep on expanding roughly up to September. Similar trend is observed in the case of leaf weight. As regards the seasonal nutrients concentration summer (April) contains maximum nitrogen and phosphorus while in winter (December) it is minimum. However, total nutrient content of leaves is minimum in April and maximum in rains (September) or autumn (November). In Tarai teak average percent retranslocation values are 45.6 % (Nitrogen) and 27.51 % (Phosphorus).

Nutrient Dynamics

Nutrient dynamics in one and thirty year old plantation system of U. P. Tarai is depicted through compartmental models (Figure 3.1, 3.2, 3.3, 3.4, 3.5 and 3.6). Standing states, annual fluxes and retranslocation are given in tanks, on shaded arrows and on unshaded arrows, respectively. In these figures direction of nutrient fluxes indicates one way movement. This flow is from soil to foliage for N and P and foliage to soil for OC.

Organic Carbon Cycling

Carbon is one of the major constituents of biomass. This is also involved in biogeochemical as well as geochemical cycling. Ranges of OC of standing state, uptake, return and retranslocation in 1 to 30 year old teak stands are 1626 - 74381, 944 - 5995, 900 - 3366 and 136 - 1377 Kg/ha/year, respectively.

Mycorrhizae and Nutrient Cycling

Mycorrhizae are the symbiotic associations between the hyphae of certain fungi and the absorbing organs - typically the roots - of plants. Mycorrhizal fungi increase the solubility of minerals in the soil, improve the uptake of nutrients (Nitrogen, Potassium and Phosphorus) for the host plant, protects the host's root from pathogens, produce plant growth hormones and move carbohydrates from one plant to another. In return they obtain carbohydrates from the plants (Hacskaylo 1972 in Alexopoulos and Mims, 1983).

Very recently the uptake of various macronutrients (Nitrogen, Phosphorus and Potassium) and micronutrients (Copper, Zinc, Iron and Manganese) in teak were studied. Mycorrhizae treated plants showed an increase in the concentration of Phosphate, Potassium and Manganese while concentration of Nitrogen, Copper, Zinc and Iron decreased. Mycorrhizae in combination with rock phosphate also gave positive results (Durga and Gupta, 1995).

Cycling of Nitrogen in
Tectona Grandis Stand

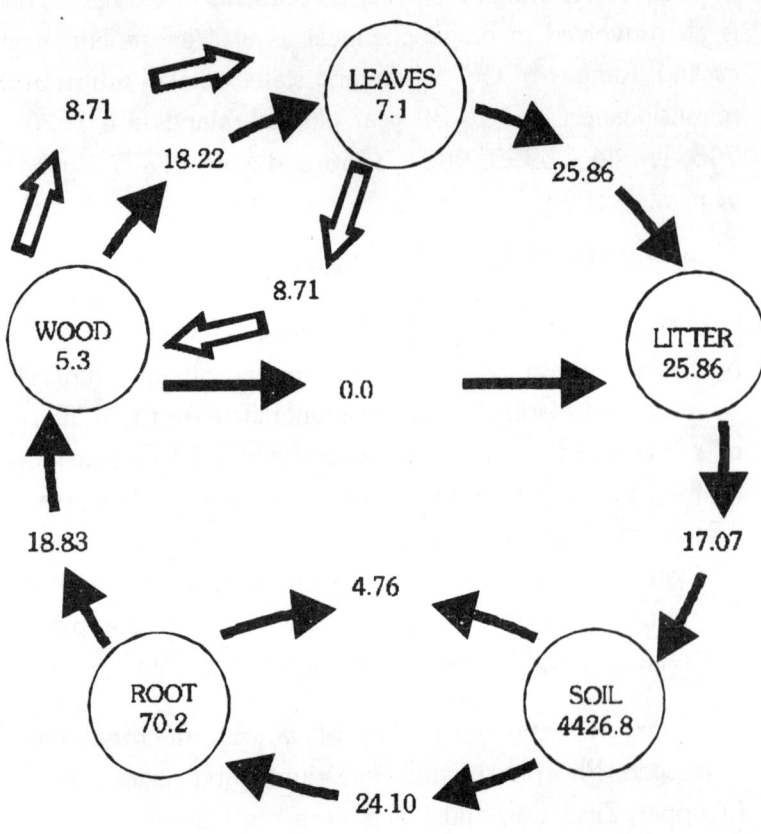

Fig. 3.1 : One Year Old Plantation

Cycling of Nitrogen in
Tectona Grandis Stand

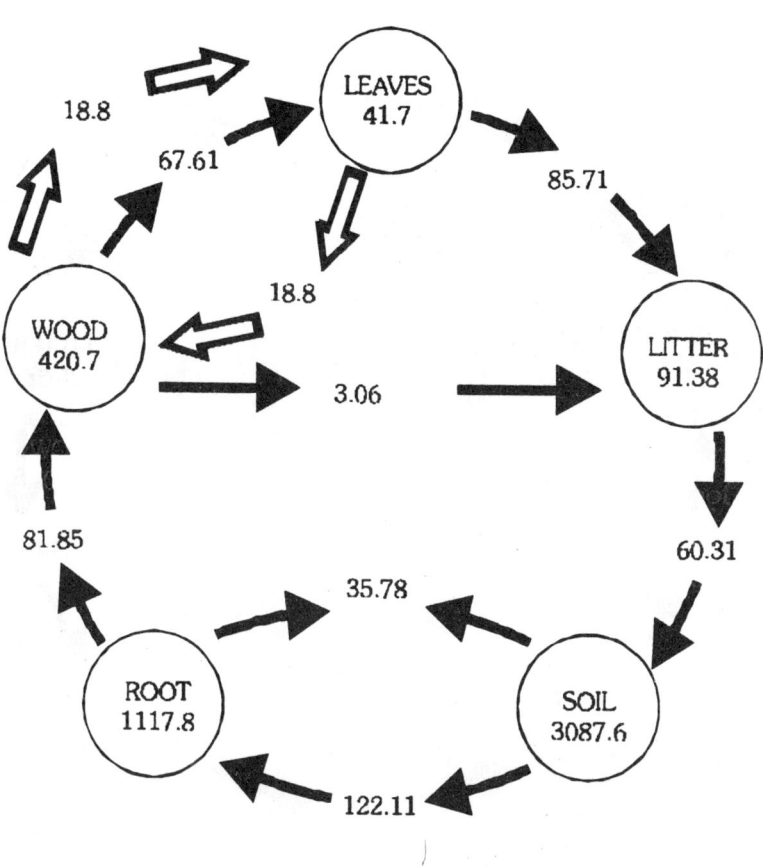

Fig. 3.2 : Thirty Year Old Plantation

Cycling of Phosphorus in
Tectona Grandis Stand

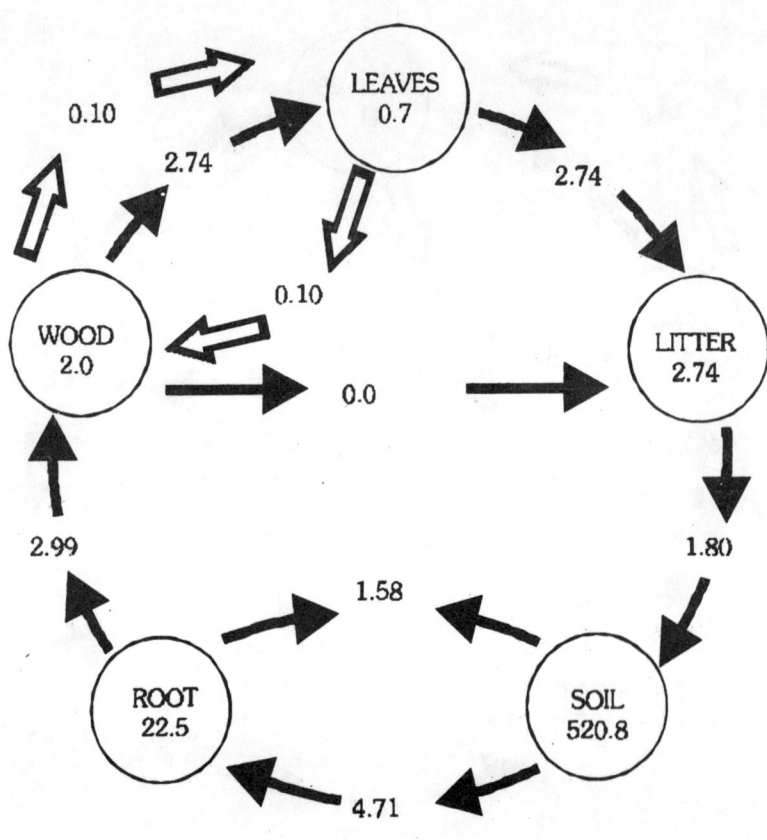

Fig. 3.3 : One Year Old Plantation

Cycling of Phosphorus in
Tectona Grandis Stand

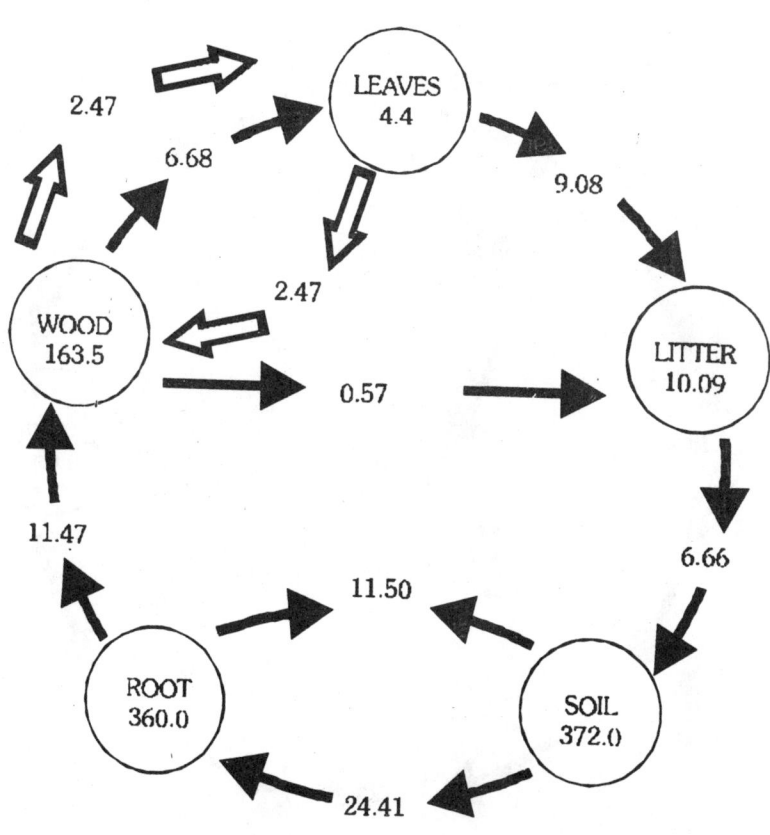

Fig. 3.4 : Thirty Year Old Plantation

Cycling of Organic Carbon in
Tectona Grandis Stand

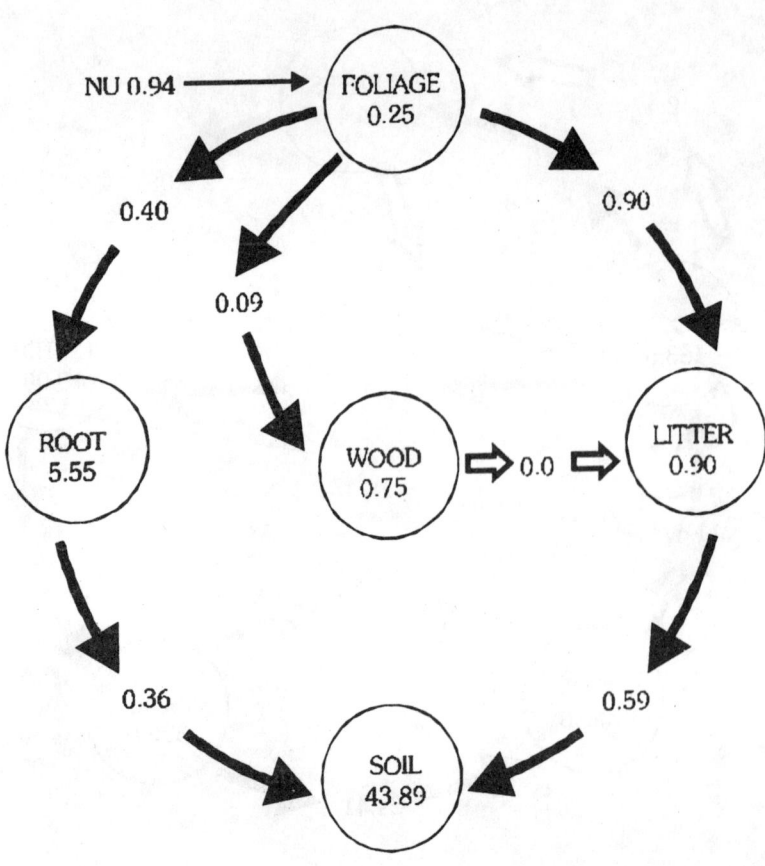

Fig. 3.5 : One Year Old Plantation

Cycling of Organic Carbon in
Tectona Grandis Stand

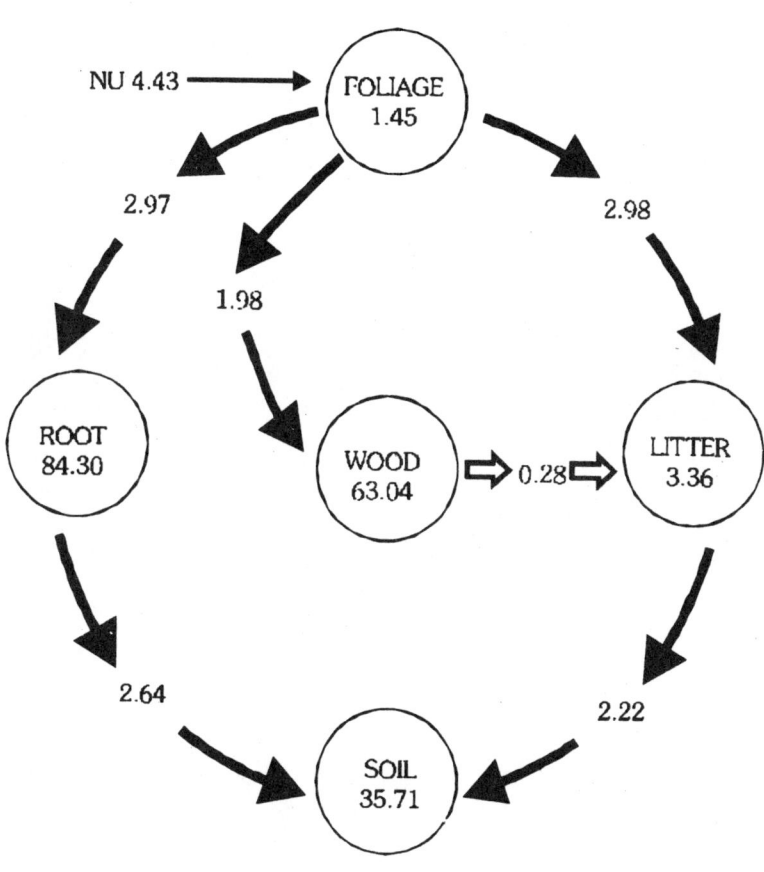

Fig. 3.6 : Thirty Year Old Plantation

Chapter 6

Plantation Technique

Chapter 6

Plantation Technique

Experience for more than a century has established that it is very easy to raise teak plantation in teak as well as non teak areas. The planting material of this species is very easily produced and propagated.

(a) Planting material

Among indigenous as well as exotic species teak is very popular as planting material. Most popular is stump planting followed by container plant and seed sowing. There are a number of reasons for this :

(i)   It can be established with relative ease and certainty.

(ii) Relatively quick establishment and growth.

(iii) Clear stem (bole) and straight form.

(iv) Hardiness and degree of immunity to damage by grazing

(v)   Relatively cheaper cost of establishment.

(vi) Standardized plantation technique.

(b) Seed collection and storage

Teak seeds profusely almost every year except for the poor seeding year. A well grown middle aged tree yields 24000 - 30000 seeds. December to March is the seed shedding period. Generally seeds are collected from the previously cleaned forest floor under the mother tree or plantation. Seeds are cleaned dried and packed in gunny bags. These bags are stacked in well ventilated cool rooms. Following precautions are required to be taken during seed collection.

(i)   Seeds should neither be collected from very young or very old trees or stands.

(ii)   It is always advisable to get seeds from known source preferably elite or candidate plus trees.

(iii)   Seeds must be dried properly before storage.

(iv)   Insect attacked or fungal infested seeds should never be stored.

(c) Pre-sowing treatment of seeds

Teak seeds are very hard and germinate with great difficulty unless its dormancy is broken by any of the following methods adopted commonly in different teak growing areas. Main purpose is to weaken the mesocarp without damaging embryo so that radicles emerge out easily.

(i)   Fire treatment - Scorching of seeds with light fire.

(ii)   Water treatment - Immersion of seeds in hot water for few hours or for several days in cold water or alternate soaking in ordinary water and drying.

(iii) Seed burial - Burial closer to anthills for varying period.

(iv) Weathering - Exposure to rain and sun for weeks or months.

(v) Pit treatment - Keeping in pits in alternate layers separated by earth and flooded with water.

(vi) Chemical treatment - Soaking in concentrated sulphuric acid for about 20 minutes and thorough washing.

(vii) Kiln treatment - Keeping in seasoning kilns for a week (temperature - 27 ° C and humidity 70%).

(viii) Cow dung treatment - Seeds soaked in cow dung slurry and spread on ground with partial shed.

(ix) Mechanical treatment - Damaging the mesocarp by the rough surface of roller fitted in drum (Bapat and Phulari, 1995).

(d) Seed sowing

Pre - treated seeds are sown either by broadcasting or in line on slightly raised nursery beds generally of 13 m x 3 m size. The usual rate of sowing is 9 - 13.5 kg seeds per bed. Straw mulching may be done to the aid of seeds. In drier areas bed shading is required. By and large teak nursery does not require regular watering, however, adequate watering improves quality of saplings if monsoon is late or uncertain. Ash and cow dung are used as fertilizer at the time of sowing.

(e) Production of planting material

(i) Entire transplants

Germinating seeds are left as it is on the nursery beds till seedlings grow up to the plantable height of 10 to 20 centimetre in two to three months and are ready for out planting. During planting they are used as bare or naked entire transplants. Entire transplants are also raised in containers like leaf cups or *Donas* and polythene bags. Germinating seeds are pricked out in these containers and grown till they are ready for planting.

(ii) Stumps

Stumps are used very commonly and successfully as replacement of entire transplants in many states. They are the 15 to 30 centimetre (20 to 25 centimeter being most common) long part of entire transplants with collar diameter of one centimetre. Root and shoot are properly pruned and sharply chopped, at top and bottom both.

(f) Planting technique

Correct selection of site is of utmost importance. Wrong selection, despite subsequent care, is almost certain to end in a disappointment. Proper timing and preplanting operations like slash burning and clearing of plantation sites are also equally important. Different method of planting are as follows.

(i) Direct sowing

This is an easy, economical, old and now almost obsolete

method of planting. This involves patch or line sowing of seeds by broadcasting or dibbling. Heavy mortality, poor growth rate and growth form resulted in rejection of this method.

(ii) Naked root planting

Entire plants with naked roots are planted in cubic pits of 30 to 45 centimetre depth at a spacing of 1.8 m x 1.8 m or 3.5 m x 3.5 m. The collar of the seedling is kept at ground level and soil is filled properly.

(iii) Container planting

Leaf cup or polythene bag plants, latter being more popular, are planted in pits at prevalent spacing.

(iv) Stump planting

This is the most successful and prevalent method of planting although a bit costly than earlier described methods. Stumps are firmly planted in holes made by crowbar or bamboo or wooden sticks. No soil preparation is required. Stumps are also planted in pits in some areas. However, experiment has proved that pits have no advantage over stake holes. Spacing of stumps vary from locality to locality and country to country. 2m x 2m spacing is more common. For high or medium rainfall areas 3m x 3m spacing is more suited.

Precautions :

(i)   Plantation sites with stiff soil and poor drainage should not be selected. Water logging for more than twenty four hours cause considerable damage to the plantation.

(ii) Lateritic, clays and black cotton soils must be avoided.

(iii) Heavily grass infested areas with *Imperata* should not be selected.

(iv) Planting schedule must be followed strictly and if unavoidable with minimum variation.

(v) Where ever feasible pure plantation should be avoided because damages arising due to it reduce the quality of timber.

Pur versus Mixed Plantation

Many disadvantages of monoculture or pure plantations raised in forest or out side have been recognised and mixed planting as the remedy has been suggested. However, certain facts regarding pure *vis a vis* mixed plantation are as follows:

(i) Stand of pure teak is often prone to attack by defoliator (Centeno, 1997). More than 280 insect pests are reported to attack teak in India and neighbouring countries. In such plantation insect attack (of chief pests - defoliator and skeltoniser) and spread is severe. Mixing of other tree crops which host the parasites of these insects could be useful in checking the spread. These tree species are *Casia fistula, Grewia tilifolia, Kydia calicina, Terminalia tomentosa, Shorea robusta* etc.

(ii) Pure teak plantations are reported to have moderating effect on soil nutrients. Possible impoverishment of soil could be replenished by mixing soil enriching species. Intercropping of

Leucaena glauca has been very useful for this purpose. In Thailand this species is used very frequently as nurse crop (Stem and Roche, 1974).

(iii) Virtually no understorey or herbaceous vegetation under pure teak supplemented with huge amount of dry litter on the ground causes fire in the plantation. This not only harms the standing crop but also clears off the humus protecting layer from splash erosion. Both crown gap filling and fire retarding can be effected by introducing crop mixture. Singh *et al,* (1985) while studying the changes in soil properties of different plantations including teak have suggested that in order to maintain natural ecological balance and avoid drastic changes in soil properties mixed plantation should be preferred to pure plantation.

Clearfelling of natural forests and creation of pure teak plantation can lead to washing away of the soluble minerals, leaving behind insoluble Silica and Sexquioxides in A - horizon. In heavy rainfall areas soil wash can remove A - horizon altogether leaving behind the B - horizon which is low in SiO_2 and rich in Al_2O_3 and Fe_2O_3. This behaves like laterite (Kadambi, 1993).

A contrary view is also on the record that for damages species purity cannot be ascribed in particular, and cannot be cured by growing mixed crop. In Java it was not found advisable to plant teak with mixture of other species because of quick crown crowding, sensitive nature of root to competition and heavy branching in mixed stands (Kadambi, 1993).

Chapter 7

Plantation Care

Chapter **7**

Plantation Care

To optimise the yield certain post plantation care like thinning, pruning, weeding etc. are to be taken up essentially because like pioneer species, teak is unable to stand much competition from other plants or from trees of the same species. It becomes more obligatory when the crop is managed on short rotation basis.

(a) Thinning operation :

Teak is a strong light demander. Congestion or suppression makes it suffer badly. As such primary concern of thinning / tending is to provide and regulate adequate growing space to the individual tree and thus to the crop as a whole to ensure their optimum development (Negi, 1996). Therefore, timely thinning is needed to foster the crown development. It has been established that early thinning is required in teak and if it is delayed beyond 10 years the crop does not respond fully to later thinning. Sarlin (1966) has also reported that teak shows marked difference in growth before and after thinning. Increment was found depressed before thinning but it resumed quickly after thinning especially in the larger trees. Thinning was done at the age of ten years. Pre and post thinning measurements were taken at eight and twelve year, respectively. Diametre, basal area and volume were selected as growth parameters. Removal of trees was roughly around fifty percent of original stocking.

Thinning intensity depends upon the growth, density and site quality of the plantation. Assuming the common density of 2000 trees per hectare at 2m x 2m spacing on good sites two thinnings for short rotation crop and few more subsequent ones are prescribed for long rotation stocks. Dhanesh Kumar *et al* (1997) have concluded that in early stage of development teak needs heavy intensity of thinning while in the later stage light thinning is required.

Following are the different thinning operations for teak plantations.

(i) First thinning

When the crop reaches top height of 8 metre, the first thinning is made. Every second tree is removed to reduce the density to 50%. This removal is staggered in alternate row. Some selection criteria may be applied to favour well formed trees or to remove badly formed ones. On good sites top height of 8m is attained in four years. Poor sites take much longer time and generally thinning is not recommended in them.

(ii) .Second thinning

This is also the semi - mechanical thinning like previous one. Top height in this case is 13 to 16 metre which is attained in seven to 12 years. Again 50% trees are removed alternately to leave behind 25% of original stocking.

(iii) Subsequent thinnings (3rd, 4th, 5th and 6th)

Third thinning is usually an ordinary thinning which consists of removing the lower canopy. It is usually done between 10 to 15 years on good areas. For poor sites it should be postponed till 40 years. At post juver.le stage when height growth starts lowering down little bit and diameter increment is quite sizable, criterion for thinning selection should be diameter related parameter for example stand basal area. Keogh (1987) has suggested that for third and subsequent thinnings standing basal area should be allowed to build up to 20 or 21 m^2/ha, when a removal of 6 m^2/ha should be made. For 3rd, 4th, 5th and 6th thinning 2 in 5 (40%), 6 in 15 (38%), 5 in 14 (36%) and 5 in 14 (36%) trees, respectively, should be removed. During these thinnings effort must be made to avoid felling of the best trees which would constitute the final crop. Varied thinning regimes in different states are given in Table 7.1.

Table 7.1 : **Different thinning regime (period in years) in different states (Source : Kadambi, 1993).**

States	First	Second	Third
Assam	5	10	15-20
Madhya pradesh	5	10	20
Tamil Nadu	3-5	6-10	10-18
Uttar pradesh	5	10	15
West Bengal	4-5	8-10	13-15

However, at poor sites mature teak trees use canopy space less efficiently. This indicates that thinning should become more intesne and frequent as the stand matures (Larson and Zaman, 1985).

(b) Weeding :

Teak is very susceptible to competition by grass and weeds for available nutrients. It can not survive without repeated weedings. Usually weeding is done in first and second year. Third year weeding is done in very few areas. Different types of weedings in practice are fork weeding, scrape weeding, weed pulling and weed cutting. First year weeding is done three times in a year. Two weedings per year are required in second year. In third year it is needed once or twice, where ever necessary.

(c) Pruning :

Branch cutting on the lower side of the bole is believed to help in clear and knotless bole formation. However, experiments in Nilambur have shown that pruning till the year of first thinning is not beneficial to teak. It is also held that pruning is of no use if it is done late in the rotation. Therefore, it is logical to do it during the gap of second and third thinning.

Teak has the propensity to produce adventitious branches and epicormic shoots next to the scars caused by pruning. To prevent their development, pruning should be carried out just after the period when most new leaves are produced (Centeno, 1997).

(d) Protection from damages :

Teak plantations suffer many kind of climatic, abiotic and biotic injuries. Climatic and abiotic factors are same almost every where but biotic factors are different in different regions. South east Asia being the native region maximum number of pathogens and insects are found in the peninsular India, the teak zone of this country. Though most of the causal factors harm teak insignificantly (Table 7.2) some serious damages are discussed below.

(i) Fire

Teak has remarkable resistance against fire but young plantations, up to the age of 10 years, fire damage can be serious. As indirect damage epidemic defoliation is induced. Though older plants are not killed growth retardation and form deformation are evident. Fire also causes secondary loss as soil erosion gets enhanced after ground vegetation burning. Regular clearance of external fire lines and controlled burning in plantation are only effective measure against fire control.

(ii) Insects

A large number of insects are reported to attack teak. But *Hybloea purea* and *Hapalia macharalis*, former a defoliator and latter a skeletonizer, are of seriously damaging nature. In both the cases of total or partial damage of leaves loss of increment is reported to the tune of 1/12 or 8.2% of total annual increment. Other than the yield timber quality loss is also evident due to development of deformities in the tree. Direct control measure like aerial dusting or spraying over vast area is not economical, however, small plantation patches can be managed easily. Prevention is possible by eradicating secondary food plants of the insect. Introduction of parasites and predators could also be tried as biological control.

(iii) Mistletoe

Loranthus longiflorus, a mistletoe, occurs under very widely varied conditions of site, aspect, slope and soil. Poor sites are most favoured. Pure teak plantations, especially from sapling to pole stages, become very common victims of this species. Lopping of infected branches is the oldest and established method of control. Selective phytocides (Copper sulphate and Femoxone) can also be tried on infested trees.

(iv) Theft

Man has become one of the greatest enemies of teak plantation these days. Since the wood is very costly and is in great demand illicit removal is very frequent in forest plantations especially located near the high population area. It is the tendency of the thieves to remove dominant trees of young plantations which ultimately affects the yield potential of the stand.

Table 7.2: **Different causal organisms of damages in teak in India (Source : Weaver, 1993).**

Causal organisms	Damages	Treatment
Cossus cadambae	Die back	*
Dichorius puntiferalis	Poor fruiting	Herbicidal
Hapalia machaeralis	Defoliation	Biological
Hyblaea purea	Defoliation	Biological
Auricularia polytricha	Wound parasite	*
Cercospora tectonae	Leaf spot	*
Corticum salmonicolor	Bark fissure	*
Marasmiellus ignobilis	Spongy trunk	*
Nectria haematococca	Bark necrosis	*
Olivia tectonae	Seedling defoliation	Fungicidal
Phialophora richardsiae	Wood decay	*
Phomopsis variosporum	Leaf fall	*
Phyllactinia guttata	Leaf necrosis	Fungicidal
Phyllosticta tectonae	Leaf spot	*
Pseudoepicoccium tectonae	Premature leaf fall	*

Pseudomonas sp.	Seedling collar rot, Wilt	*
Scleratiom rolfsii	Wilting, spotted leaves	*
Uncinula tectonae	Moisture loss, death	Chemical
Peniphora rhizomorpha-sulphurea	Root rot	*
Rigidoporous zonalis	Root, heart rot	*
Bjerkandera adusta	White rot	*
Flavodon flavous	White rot	*
Fomes lividus	Rot in low cut stems	*
Ganoderma applantum	White rot	*
Phellinus lamaoensis	Brown rot	*
Polyporus rubidus	Brown rot	*
P. Shoreae	Patridgewood	*
Dendrophthe falcata	Parasite	Chemical
Loranthus longiflorus	Parasite	Chemical, Lopping

* not recognized

PLATE 2 Tending effect in Teak

Unthinned Teak plantation

Thinned Teak plantation

.

Chapter 8

Growth and Yield

Chapter 8

Growth and Yield

eak is moderately fast growing species. It grows faster in initial years but slows down afterwards (Dhanesh Kumar *et al*, 1997; Jha, 1995; Negi, 1996). Growth is generally measured as diameter, height and basal area of the crop at particular time/age. A comparative statement of age, basal area and density presented in Table 8.1 gives an idea of teak growth in different regions inside as well as outside India.

Yield of teak generally depends on rotation, harvesting cycle, of the crop which varies from forest types to management systems. In most of the areas where teak occurs in mixed stands and the crop is almost even aged rotation ranges from 70 to 150 years. In coppice or coppice with standard system, generally in drier localities, it is 40 to 60 years. The plantation crops are managed at 50 to 80 years. Recently harvesting cycle has drastically been reduced to 20 to 30 years with coming up of green gold venture emphasizing on agrocommercial farming of teak. In tropical America also most of the plantations are managed with this far shorter rotations (Centeno, 1997).

Table 8.1 : **Comparative statement of basal area, age and density in different teak growing regions.**

Location	Age years	Basal area sq m / ha	Density t/ha/yr	Author(s)
India				
Gorakhpur	5	15.3	2068	Faruqui (1972)
	14	32.6	1022	
	30	54.9	682	
Dehradun.	33	-	630	Seth & Kaul (1978)
Varanasi				
Chakia	4	5.1	3490	Karmacharya and
	14	7.0	1040	Singh (1992)
	30	11.0	474	
Chandraprabha	15	4.1	467	Singh & Mishra (1979)
Surguja	10	11.5	-	Sharma & Naik
	14	8.7	-	(1989)
	22	20.1	-	
Bijawar	20-23	9.9	1528	Chaubey et al
Kalpi	20-23	11.0	1678	(1988)
Haldwani	1	1.2	1183	Jha (1995)
	5	5.9	1728	

Location	Age years	Basal area sq m / ha	Density t/ha/yr	Author(s)
	11	10.57	376	
	18	12.83	512	
	24	16.33	273	
	30	18.52	323	
Kurseong Pashok	18	20.18	1000	Gangopadhyay
Bamanpukuri	24	32.68	800	et al (1987)
Reyang	18	34.06	1500	
	47	90.8	1000	
Puerto Rico	24	20.9	955	Weaver and
	37	19.7	674	Francis
	40*	30.3	382	(1992)
	44*	37.8	509	
	45*	39.7	487	
	47*	46.2	525	
	49	25.9	732	
	51	26.6	469	

Location	Age years	Basal area sq m / ha	Density t/ha/yr	Author(s)
Venezuela	4	7.6 *	1308	Hase and Foelster
	6	13.5	1208	(1983)
	8	19.1	1275	
	9	38.0	850	

* High rainfall area (>2000mm)

However, yield estimation, generally dependent on growth parameters, has more relevant scope for the brevity of this text. Commercial measure of the tree is merchantable volume of timber. But ecological measure which is getting more and more attention, currently, due to various end uses of tree parts other than the bole is biomass and productivity. Following paragraphs embody the details of these parameters.

(a) Volume production

Depending on site quality and exploitable age volume production of teak timber (exclusive of bark to 20 centimetre top diameter over bark) varies from 0 to 495.5 cubic meter per hectare (Table 8.2). Haldwani is reported to be one of the best teak growing regions of India (Reddy, 1995). Recent performance recorded by Jha and Singh (1997) in this area shows slightly better picture than as reported earlier by Kadambi (1993). Please refer to Table 8.5 also.

Table 8.2 : Yield of plantation teak in India (Volume in cubic metre per hectare; Source : Kadambi, 1993).

Age in years	Site quality*			
	I	II	III	IV
20	68.5	27.6	6.2	0
30	171.6	78.8	17.3	0.7
50	351.2	189.6	58.4	6.2
60	412.8	239.1	92.1	14.9
80	495.5	312.8	172.3	46.4

* Top height in feet : I;120-100, II;100-80, III;80-60, IV;60-40.

(b) Biomass Estimation

Huge amount of energy and organics are stored in a forest ecosystem as biomass and soil organic matters. Estimates of biomass is essential for determining the status and flux of biological materials in ecosystem and also for understanding the dynamics of ecosystem (Satoo,1970; Anderson, 1970). Above ground biomass of teak recorded between one to thirty eight years at different sites varies between 1.9 to 382 t/ha (Table 8.3).

Table 8.3 : Biomass of teak plantations in different teak growing areas.

Location	Age (year)	Above ground biomass* t/ha	References
Dehradun, U.P.	38	129.58	Kaul et al (1979)
Teliamara, Tripura	20	114.37	Negi et al (1990)
Raipur,M.P.	30	11.34	Singh & Gupta (1993)
Gorakhpur, U.P.	5	49.6	Faruqui (1972)
	14	158.9	
	30	301.9	
Dehradun, U.P.	33	86.5	Seth & Kaul (1978)
Varanasi,U.P.	4	25.23	Karmacharya & Singh (1992)
	14	39.9	
	30	76.0	
Coimbatore, T.N.	20	148.8	George & Verghese (1992)
Haldwani, U.P.	10	74.6 (90.0)	Negi et al (1995)
	20	90.7 (108.6)	

Location	Age (year)	Above ground biomass* t/ha	References
	30	164.1 (192.6)	Jha (1995)
	1	1.9 (3.1)	
	5	26.7 (30.5)	
	11	59.9 (68.1)	
	18	65.9 (75.0)	
	24	104.4 (118.4)	
	30	123.3 (141.1)	
Llanos, Venezuela	4	42.5	Hase & Foelster (1983)
	6	70.5	
	7	116.0	
	9	194.0	
Gambari, Nigeria	18	225 - 382	Ola - Adams (1993)

* Figures in parentheses are total biomass

Within Uttar Pradesh on account of biomass production Gorakhpur seems to be the best suited area followed by Haldwani, Dehradun and Varanasi. Available data outside Uttar Pradesh indicates that Coimbatore, Tamil Nadu is much better area than Raipur, Madhya Pradesh. However outside India Llanos, Venezuela and Gambari, Nigeria have higher yields.

(c) Annual Productivity

Annual productivity, also expressed in net primary productivity, is the energy source that sustains life of all heterotrophs and maintains the composition and balance of an ecosystem. It has become impertinent to recognise productivity as the basis of forest land management. In teak biomass accumulation ratio, basal area and density are directly proportional to annual productivity. Roughly around forty percent of annual biomass production is retained in the form of annual standing state and sixty percent is released in the form of litter. Comparative annual productivity of different teak sites are given in Table 8.4. At different location rate of annual productivity is highly variable. Major reasons for this must be the variable climate and response of teak tree to it at different age.

Production in Moist Region

(i) In Uttar Pradesh

Biomass, annual productivity and volume in an age series plantations of Kumaun Himalayan Tarai are given in Table 8.5. In Tarai teak total biomass at different ages e.g., 1, 5, 11, 18, 24 and 30 years are 3.101, 30.330, 68.172, 75.014, 118.423 and 141.345 t/ha, respectively. In these stands total biomass as well as above and below ground biomass increased with increase in age. Biomass share of different tree components at different ages are depicted in Figures 8.1, 8.2, 8.3, 8.4, 8.5 and 8.6.

Table 8.4 : Net annual productivity of teak in different regions.

Location	Age years	Basal area m² / ha	NAP t/ha/yr	Author(s)
India				
Gorakhpur	5	15.3	12.3	Faruqui (1972)
	14	32.6	11.0	
	30	54.9	12.0	
Dehradun	33	-	8.4	Seth & Kaul (1978)
Varanasi				
Chakia	4	5.1	25.6	Karmacharya & Singh
	14	7.0	14.0	(1992)
	30	11.0	12.9	
Chandraprabha	15	4.1	3.9	Singh & Mishra (1979)
Surguja	10	11.5	5.3	Sharma & Naik (1989)

Location	Age years	Basal area m² / ha	NAP t/ha/yr	Author(s)
Haldwani	14	8.7	4.7	Jha (1995)
	22	20.1	11.6	
	1	1.2	1.78	
	5	5.9	13.39	
	11	10.57	10.51	
	18	12.83	10.41	
	24	16.33	6.53	
	30	18.52	10.35	
Thailand	40 - 80	-	8 -10	Centeno (1996)

However, percentage contribution of above ground (87.24 - 88.25 %) and below ground (12.75 - 11.75 %) biomass among the ages do not show much variation.

Unlike biomass relationship of annual productivity with age is unstable. This is dependent on density as well as changed growth response of this species at different age. However volume production has direct relationship with increasing age.

Linear regression equations for volume and biomass are also presented in Table 8.6 and 8.7. These equations are highly significant. They may be utilised in determining the growing stock in the area by non destructive method.

Table 8.5 : Volume, Biomass and Productivity of various teak stands (Source : Jha 1995; Jha and Singh, 1997).

Age (year)	Volume (m³/ha) Bole**	Bark	Biomass* (t/ha)	Productivity (t/ha/yr)
1	4.14	1.78	1.42	2.08
5	54.07	17.20	18.29	13.39
11	99.63	16.24	47.24	10.51
18	126.09	21.42	68.99	10.41
24	164.95	24.53	76.75	6.53
30	210.15	26.77	100.54	10.35

* Above ground, ** Over bark

Table 8.6 : **Linear regression equations between Diameter (x) and volume (y) in different aged teak stands [y=a+bx].**

Age (year)	Compo- nents	Cons- tant(a)	Slope (b)	r^2
18	OB	-0.3142	0.0309	0.9586 **
	UB	-0.2958	0.0277	0.9425 **
	BAK	-0.0183	0.0031	0.8521 **
24	OB	-0.5071	0.0432	0.9186 **
	UB	-0.4613	0.0381	0.9072 **
	BAK	-0.0458	0.0051	0.6431 **
30	OB	-0.6266	0.0474	0.9061 **
	UB	-0.5782	0.0426	0.9045 **
	BAK	-0.0484	0.0048	0.7683 **
1-30	OB	-0.3293	0.0362	0.8874 **
	UB	-0.3026	0.0321	0.8799 **
	BAK	-0.0267	0.0041	0.7887 **

OB = volume over bark, UB = volume under bark, BAK = bark, N = not significant, * = significant at 5%, ** = significant at 1%.

Table 8.7 : Linear regression relationship (y = a + b . x) between biomass of tree components (y, t/ha) and dbh (x, cm) for *Tectona grandis* age series plantation.

Age (years)	Components	Intercept (a)	Slope (b)	r^2	Std Err Coef
30	Bark	-22.036	1.849	0.792**	0.236
	Bole	-267.819	20.481	0.868**	1.995
	Branch	-80.007	4.239	0.514**	1.029
	Twig	-41.834	2.685	0.690**	0.449
	Foliage	-3.015	0.432	0.367**	0.141
	Above ground	-415.838	29.740	0.910**	2.324
	Stump root	-48.297	3.029	0.692**	0.505
	Lateral root	-45.447	2.547	0.717**	0.399

Age (years)	Components	Intercept (a)	Slope (b)	r^2	Std Err Coef
	Below ground	-93.744	5.576	0.808**	0.677
	Total	-509.583	35.317	0.935**	2.318
24	Bark	-16.389	1.687	0.942**	0.131
	Bole	-237.146	19.205	0.903**	1.986
	Branch	-18.229	1.935	0.445*	0.682
	Twig	-19.665	1.862	0.734**	0.353
	Foliage	-0.351	0.182	0.366*	0.760
	Above ground	-291.857	24.879	0.921**	2.292
	Stump root	-8.399	1.444	0.902**	0.150
	Lateral root	-15.527	1.311	0.694**	0.274
	Below ground	-23.927	2.755	0.905**	0.281

Age (years)	Components	Intercept (a)	Slope (b)	r²	Std Err Coef
	Total	-315.784	27.634	0.922**	2.526
18	Bark	-13.726	1.520	0.964**	0.103
	Bole	-152.007	13.647	0.961**	0.963
	Branch	-16.392	1.361	0.751**	0.277
	Twig	-28.101	2.206	0.763**	0.434
	Foliage	-10.280	0.824	0.840**	0.127
	Above ground	-220.991	19.590	0.958***	1.437
	Stump root	-19.717	1.681	0.844**	0.254
	Lateral root	-21.759	1.638	0.757***	0.327
	Below ground	-41.476	3.319	0.809**	0.568
	Total	-262.467	22.910	0.947**	1.908

(ii) In West Bengal

Production of teak timber in moist region of West Bengal is little higher than Uttar Pradesh. D^2H (D = Diameter, H = Height), an indicator of volume, value of 18 and 24 years old teak plantation of Darjeeling district are 0.578 m^3 (D = 17cm; H = 20m) and 1.85 m^3 (D = 22.8 cm; H = 22.8 m), respecively. These values are higher than 0.647 m^3 and 1.391m^3 recorded for same aged plantation at Haldwani. (Gangopadhyay *et al,* 1987; Jha and Singh, 1997). Apparent reasons seem to be density and climatic factors.

Biomass Distribution (1 Year)

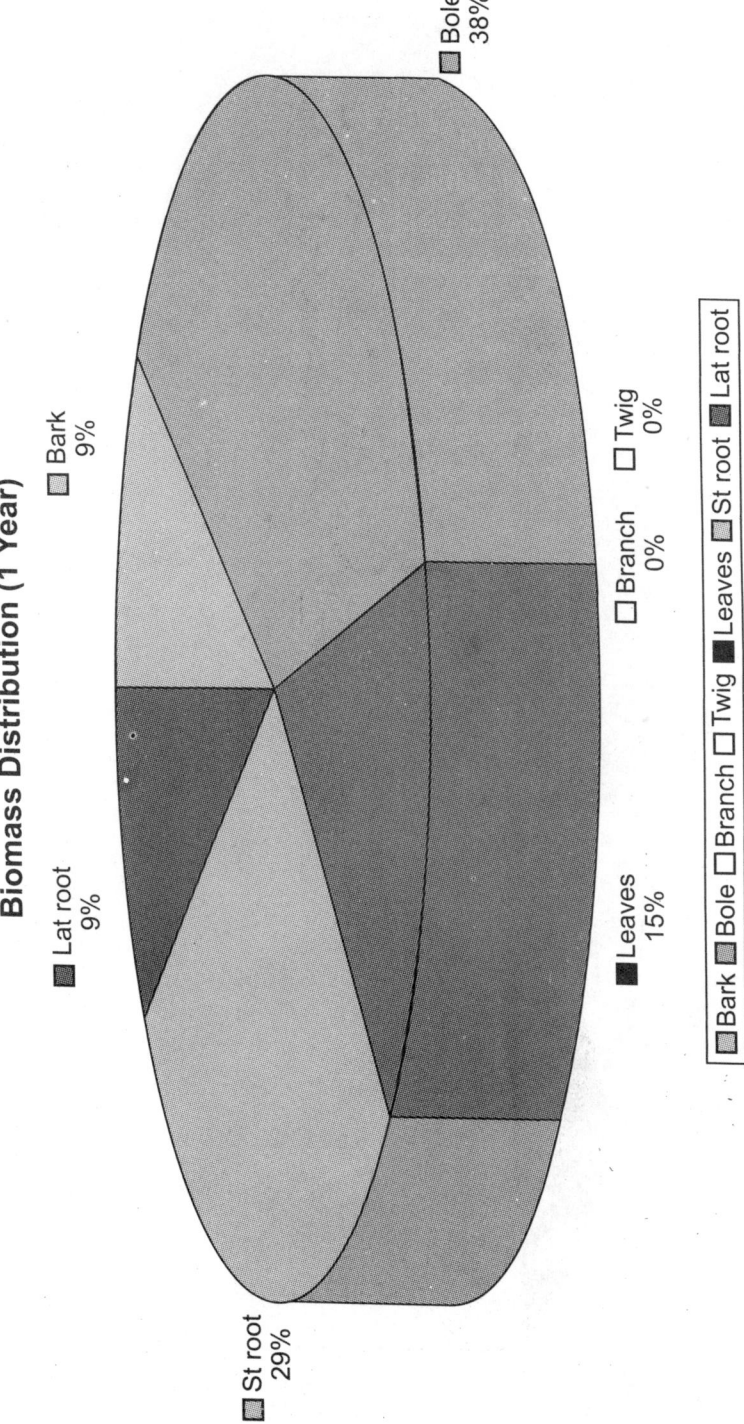

Bark 9%

Lat root 9%

Bole 38%

Branch 0%

Twig 0%

St root 29%

Leaves 15%

■ Bark ▨ Bole □ Branch □ Twig ■ Leaves ▨ St root ■ Lat root

Fig. 8.1

Biomass Distribution (5 Year)

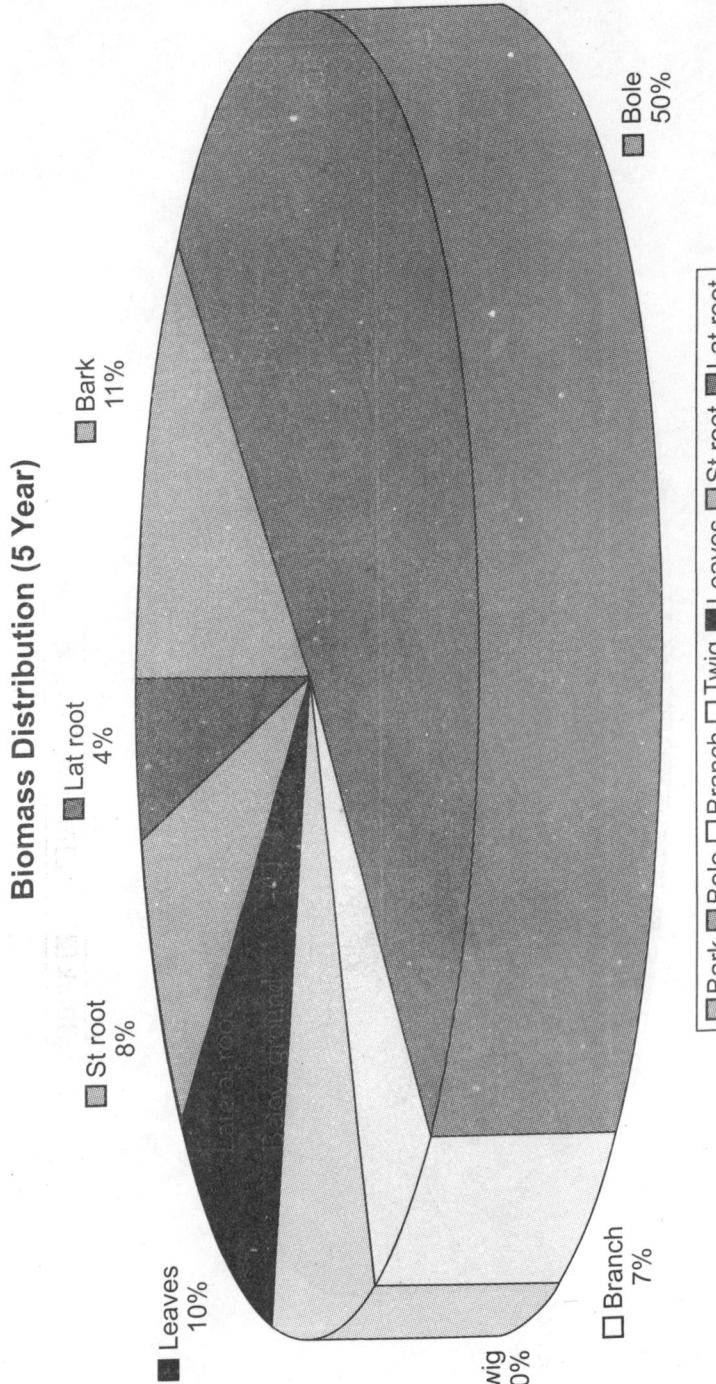

Leaves
10%

St root
8%

Lat root
4%

Bark
11%

Bole
50%

Twig
10%

Branch
7%

■ Bark ▨ Bole □ Branch □ Twig ■ Leaves ▨ St root ▨ Lat root

Fig. 8.2

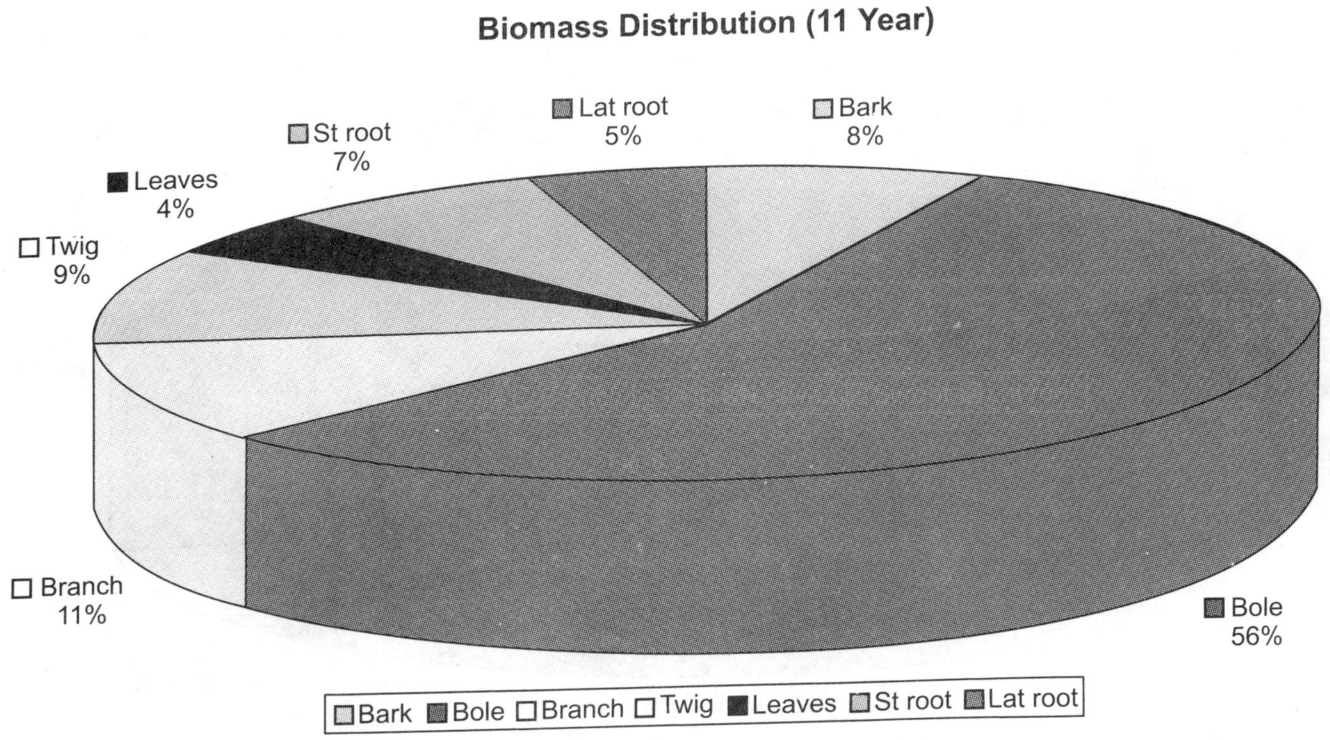

Biomass Distribution (11 Year)

Leaves 4%

Twig 9%

St root 7%

Lat root 5%

Bark 8%

Branch 11%

Bole 56%

Bark ■Bole □Branch □Twig ■Leaves ■St root ■Lat root

Fig. 8.3

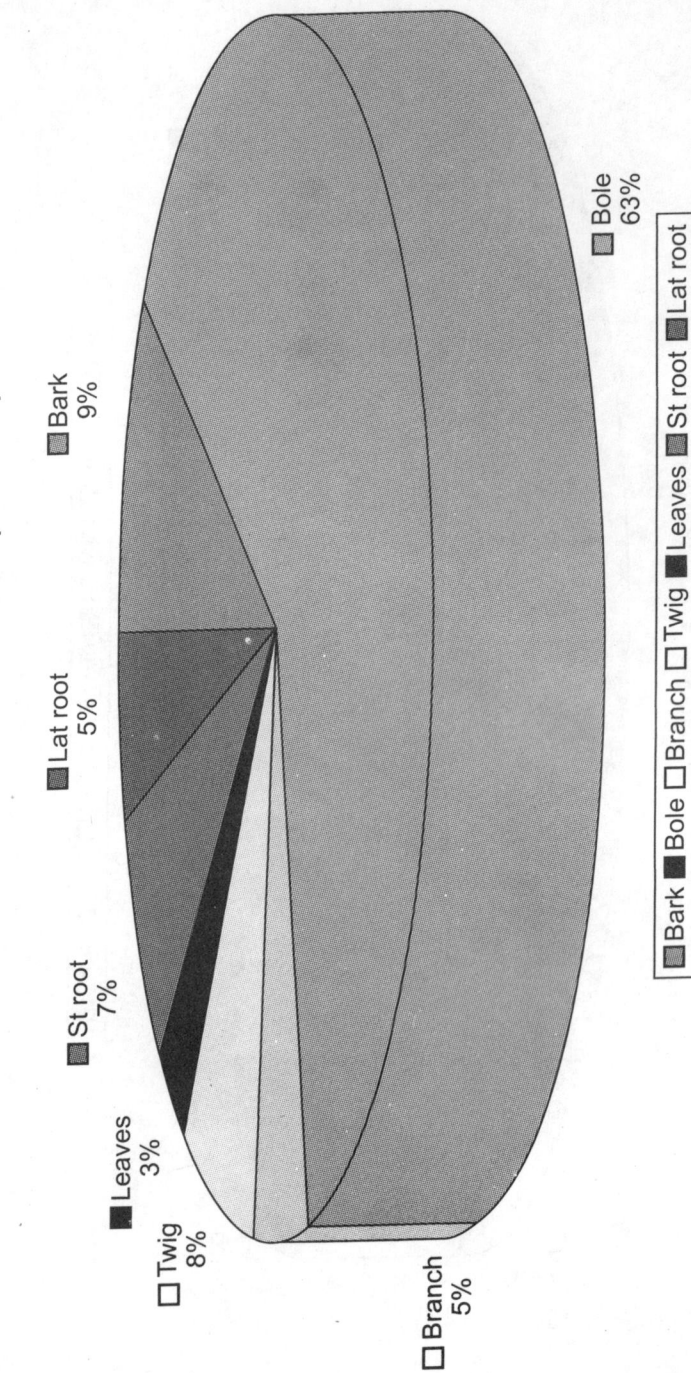

Biomass Distribution (18 Years)

Bole 63%

Bark 9%

Lat root 5%

St root 7%

Leaves 3%

Twig 8%

Branch 5%

Bark Bole Branch Twig Leaves St root Lat root

Fig. 8.4

Biomass Distribution (24 Years)

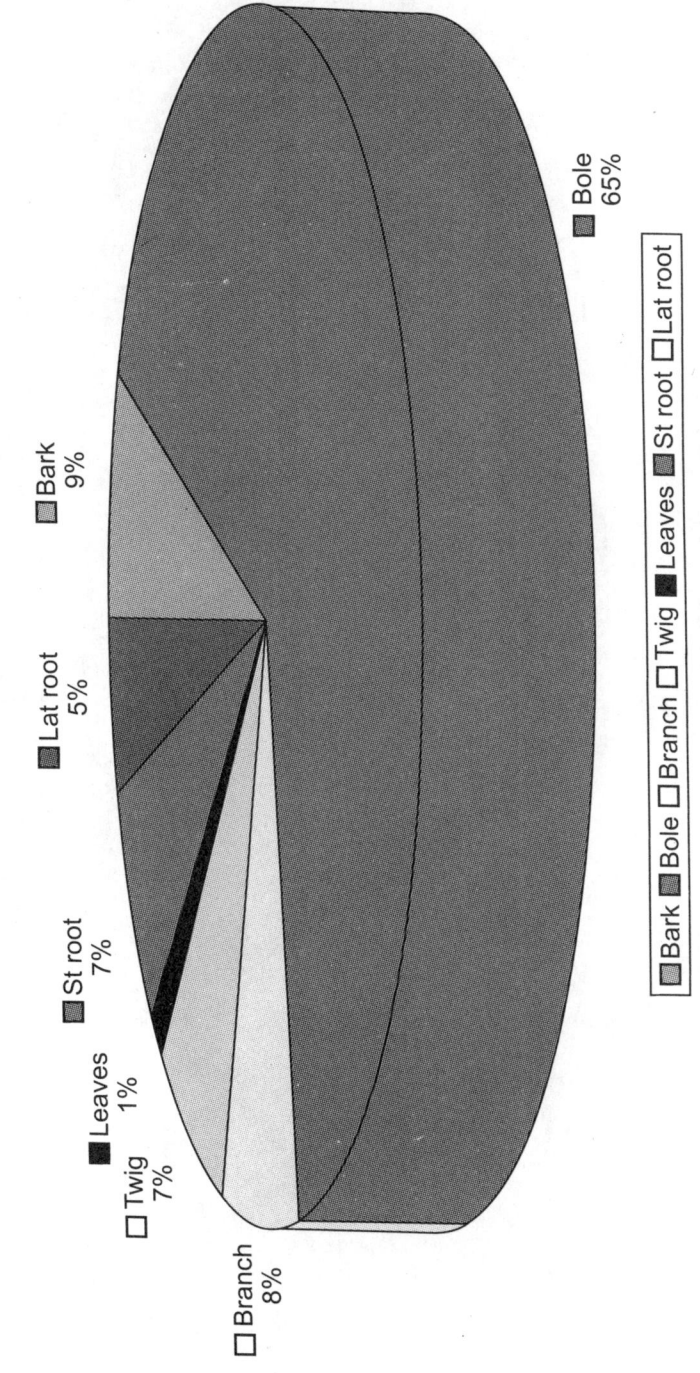

Fig. 8.5

Bole
65%

Bark
9%

Lat root
5%

St root
7%

Leaves
1%

Twig
7%

Branch
8%

Bark ■ Bole □ Branch □ Twig ■ Leaves ■ St root □ Lat root

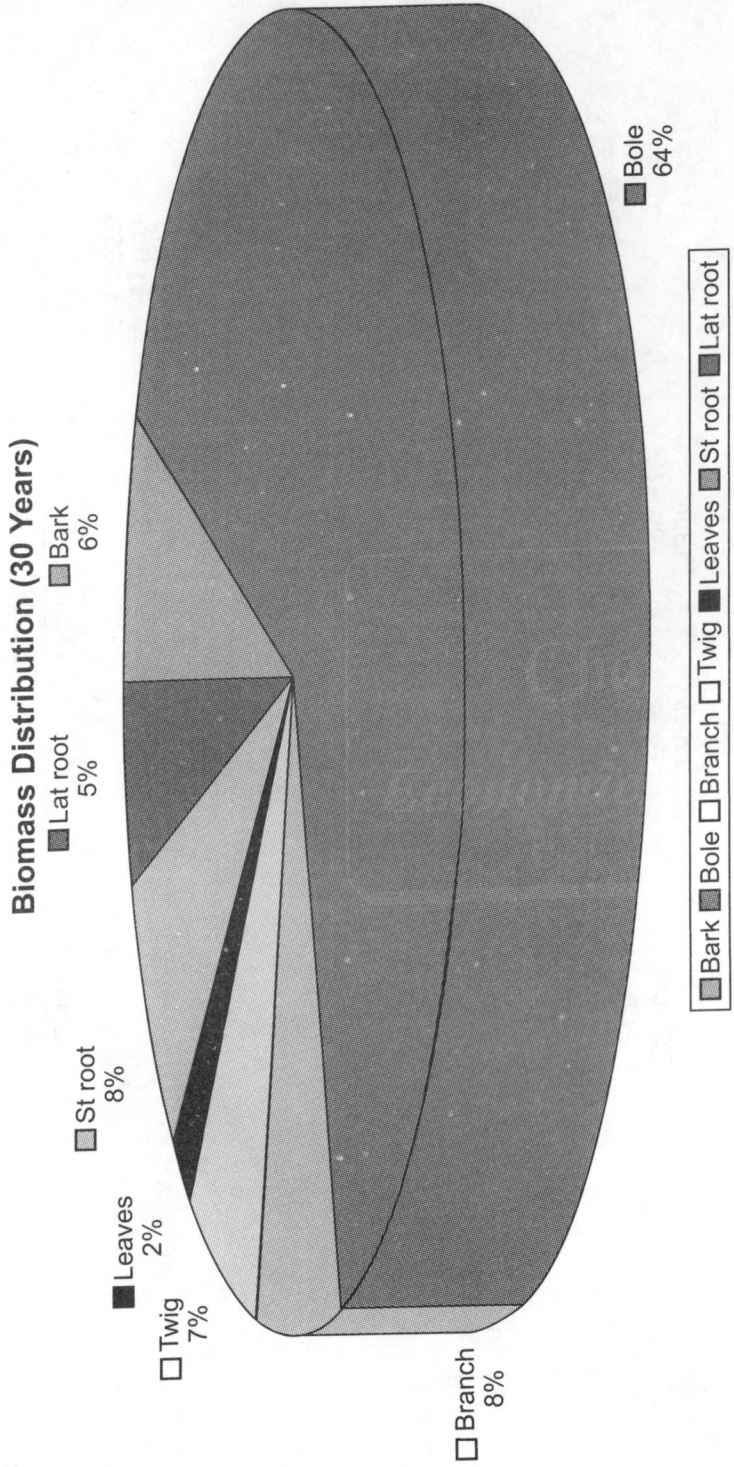

Biomass Distribution (30 Years)

Bark 6%
Lat root 5%
St root 8%
Leaves 2%
Twig 7%
Branch 8%
Bole 64%

■ Bark ■ Bole □ Branch □ Twig ■ Leaves ■ St root ■ Lat root

Fig. 8.6

Chapter 9

Economics and Marketing

Chapter 9

Economics and Marketing

s discussed earlier the financial models presented by the private companies are to be revisualised since the financial returns are based on technical feasibilities. Various results put forward show that it is not possible to fill the huge gap of recorded and projected yield of teak (Chapter 4) in short span of twenty or thirty years. However, steep rise in market price of teak wood (seventy times in thirty five years in one instance) shows enough promise in taking up teak as farm forestry or agroforestry.

Plantation economics

Economic analysis of pure teak plantation in Maharastra and Madhya Pradesh at quality I and II sites have shown positive benefit and cost ratio (3.98 - 1.96 at 16.2 - 12 % Internal Rate of Return) at younger as well as older rotation age of teak. Mixed teak plantation with Semal and Bamboo at the age of fifty years also showed healthy B/C ratio (2.97 and 2.63, respectively) at 13.8 % IRR (Tewari, 1993).

Certain assumptions for the viability of the project are as follows:

(1) Plantation should be established with all tending cares.

(2) In initial years agroforestry should be practiced.

(3) Timely thinning operations must be carried out.

(4) Rotation must be fixed on or after twenty five years.

A model of teak management under commercial forestry in UP plains indicates that teak plantation is capable of returning its capital with interests in ten years at the rate of 16% compound interest. Of course, model is based on certain logical and feasible assumptions (Srivastava, 1997).

Marketing

Being one of the earliest and world wide recognised timber it has very good markets through out the country. Almost every state has several distinguished centres for this timber. Few important markets and price of converted timber are given in Table 9.1 (Source : Negi, 1996).

Table 9.1 : Average price (Rupees per cubic meter)
of teak wood at different major centres
(base year 1995).

Market / Centre	Round logs	Sawn timber
Ahmedabad	26500	*
Bangalore	37000	29750 - 59000
Calcutta	33000	31770 - 37065
Calicut	34500	35300 - 38850
Delhi	*	32609 - 38090
Jabalpur	29800	15250 - 38841
Jaipur	24450	15000 - 32500
Madras	22500	31800 - 42400
Nagpur	20300	14350 - 29500
Raipur	20700	*
Ranchi	21550	*

* Not available

In Uttar Pradesh Haldwani, Dehradun, Jhansi, Khatima, Ramnagar and Saharanpur are the distinguished teak marketing centres. Average sale price of UP Forest Corporation in 1997 was Rs. 11000/- per cubic metre.

Chapter 10

Crop Improvement

Chapter **10**

Crop Improvement

*T*he common practice is to collect seeds either from unrecognised source or from the forest floor randomly. Also the seedlings growing in the forest are picked up in bulk to use them in preparing stumps or direct planting. These practices definitely need improvement to secure use of quality seeds. Polythene bag plant raising also needs improvement in container size to cater the need of vigorously growing root and also in earth based potting mix to give better growing and nourishing medium.

(a) Use of Improved Planting Material

It is estimated that yield of plantation stocked from clonal orchard will increase by minimum of 20% and from selected candidate plus trees by about 15% (Anon, 1997). Therefore, quality controlled seeds obtainable from the recognised institutes must be used.

Tissue culture is the latest biotechnological method which can be utilised to propagate teak clones of high yield potential. Many institutes like Forest Research Institute, Dehradun, National Chemical Laboratory, Pune, Tata Energy Research Institute, Delhi, National Botanical Research Institute, Lucknow etc. have pioneered in developing tissue cultured clones of forestry species.

(b) Adoption of Modern Nursery Technology

Quality seedling production has direct bearing on the health and growth of plantations. Therefore modern nursery technologies which include introduction of root trainers, use of compost based potting mix and chemical fertilizers, improved irrigation (spray) system, strict monitoring etc. must be adopted. Uttar Pradesh Forest Department has ventured into the modern nursery technique on large scale.

(c) Use of Damage Resistant Varieties

Now hardy races to mistletoe are available. Some of the defoliator resistant varieties have also been recognised. Use of these varieties may help in enhanced production of timber.

(d) Use of Mycorrhizae

Mycorrhizae are known to enhance nutrient uptake of forestry species. This results in greater accumulation of resources and ultimately biomass production. Infertile soils are made efficient nutrient supplier to the plants if mycorrhizae are added to it. Therefore, they may be used as biofertilizer. There are reports of enhanced dry weight of many broad leaved species, even upto 50% or more, when grown under mycorrhizal inoculation (Kandaswamy *et al*, 1988). In teak seedlings also inoculation of mixed arbuscular mycorrhizal fungi showed significantly enhanced root and shoot biomass and percent infection in nursery. Percent infection is positively correlated with biomass production. The trees with better genome from silviculture point of view (Plus Trees) were also found to be better adapted with arbuscular mycorrhizae and showed more percent root infection than other normal trees (Verma and Jamaluddin, 1995). Seasonal variation in percent root infection is also reported in South Indian teak. High root colonization by mycorrhizae is in summer (84%) followed by late winter (62%). Early winter followed by rainy season showed minimum colonization (Raman and Gopinathans, 1992).

In spite of the ensured crop improvement through candidate plus tree (CPT) and clonal orchard seed origin seedling production has the limitation of possible genetic variation that might result in non exacting and unhealthy offspring. To get rid of this shortcoming established method of vegetative reproduction should be adopted for raising plants. This ensures improved genetic gain and its maintenance through several generations. Next chapter is devoted to vegetative propagation.

Chapter 11

Vegetative Propagation

Chapter  11

Vegetative Propagation

Forest tree species in which small genetic variation is not of much importance are mostly propagated by seeds due to so many advantages of this method in comparison to other available vegetative propagation (VP) methods. The reasons assigned are as below.

(1) Production of large number of plants in reasonable time.

(2) It is a low cost production method.

(3) It helps in introduction of new cultivar.

(4) Collection and transport of seedling is comparafively easier.

(5) Prevention from virus disease transmission.

However, VP matters a lot when genetic variation is not desired. In fact different methods of VP ensures the maintenance of genetic make up and health status of the parent stock. Some of the advantages of VP are listed as follows.

(1)   Reproduction of exact genetic composition.

(2)   High degree of crop uniformity.

(3)   Overcome complex seed dormancy, germination and viability problems.

(4)   Perpetuate both pest and disease resistance.

(5)   Mass production in quicker time.

With all these above mentioned advantages VP has a great potential for tree improvement in forestry species. It finds immense use in raising clonal seed orchards and clonal bank plantations for production of high quality seed and for speeding up tree breeding work. This method is gaining momentum fast due to its application for mass multiplication of clonal material for raising plantation of elite trees (Khullar *et al,* 1991).

VP is such an effective tool that can also be applied in improvement of *Tectona grandis.* Some of the common VP methods are discussed below in context of this species.

(i)   Cuttings from Stem and Root

Stem and root cuttings, derived from shoot and root, respectively, are the vegetative segments of suitable size. Each of the segments contains lateral and / or terminal buds. This method is simple, rapid and inexpensive.

Stem cutting : According to nature of wood cuttings are categorized as hard wood cuttings, semi hard wood cuttings and soft wood cuttings. They are further classified as easy to root, difficult to root and obstinate to root depending on rooting behaviour of cuttings. Teak comes in difficult to root category. An experiment revealed that only 20% of cuttings sprouted successfully. The cuttings were taken from 15 year old teak tree. They were 30 cm long and 4 cm in diameter. The roots were examined after 180 days. They were found healthy and vigorous and were of varied length from 20 to 26 cm (Lahiri, 1974). Seeing the limited practical utility this method is not adopted to propagate this species. However, rooting problems of certain species have been solved by the use of rooting hormones.

Rooting hormones : They are synthetic substances applied to the base of cuttings to aid rooting. They are derivatives of growth substances or regulators that occur naturally in living plants. Advantages of rooting hormones are described as below (Macdonald, 1986) :

(1)   Hasten root initiation.

(2)   Increase the number and quality of roots.

(3)   Encourage the uniformity of rooting.

(4)   Increase over all rooting percentage.

Branch cuttings of teak had been successfully rooted in the nursery with the help of growth hormone. Cuttings with 1.5 - 2 cm diameter and about 20 cm long, taken from one year old branches rooted well when treated with auxins (IBA). Without auxins cuttings did not root at all or in exceptional cases rooted very poorly (Puri and Bangarwa, 1990).

Root cutting : VP by root cuttings is very simple and is generally used for succulents. However, initial success in hardwood species like teak have also been reported (Khullar *et al*, 1991).

(ii) Layering

In this method of VP root formation is initiated on the shoot by girdling and covering the branch. After root formation rooted branch is detached and planted for its establishment into a new individual. Different type of layering are Air layering, Mound layering, Trench layering, French layering, Tip layering, Serpentine layering etc. In air layering layered branch hangs in air and girdled part is enveloped with rooting medium. Other layering are modifications of air layering. In forestry and hard wood species generally air layering is adopted. Since rooting is not easy in teak this method may not be useful for this species, however, with the help of root initiators or rooting hormone success can be achieved. Research studies are needed in this field.

(iii) Grafting and Budding

It is a technique of insertion of a part of plant, generally twig of superior genotype to be propagated, in another plant in such a way that two unite and the combination develops into a successful plant. In this union former is known as scion and latter the stock. Scion contributes the shoot while stock the root part of a grafted plant. Grafting has been very successful in teak and has been adopted widely where skilled labour is not a constraint and limited number of seedlings are required. When the scion is bud instead of twigs or branches the method of VP is known as budding.

Grafting (cleft) and budding (patch) are highly successful methods of clonal propagation of teak provided that operation is carried out at 21.3 - 26.3 °C temperature and 45.5 - 62 % humidity. April and May were found more suitable period to obtain maximum success in Dehradun field conditions. The percentage success can be increased by use of Greenhouse or Mist chamber (Uniyal and Rawat, 1985).

(iv) Micropropagation

It is a specialised method of VP in which very small pieces of plant tissue are regenerated in an artificial medium under sterile conditions. It embraces the regeneration from shoot and root tips, callus tissue, leaves, seed embryos, anthers and even single cell. In woody plants shoot tips are most commonly cultured (Macdonald, 1986). Although there are limitations like Heavy initial capital investment, Employment of highly skilled personnel etc. to this biotechnology widely known as Tissue Culture it has immense advantages.

Advantages : Micropropagation has many applications summarized as under (Macdonald, 1986) :

(1) As a means to remove viruses from plants.

(2) As a very effective and rapid method to multiply clonal material.

(3) As a means of propagating all year rather than

being confined to the seasonal variations common to conventional techniques.

(4) As a particularly convenient way to export plants to other parts of world.

(5) As a means of radically reducing both the number of stock plants and the size of stock plant beds.

(6) As a means of significantly reducing the actual size of propagation area.

(7) As a means of providing higher quality, disease free saleable product.

(8) As a specialized method of raising plants from seed (cotyledons).

Several forestry species have been multiplied by this method and improved varieties have been developed. Clonal multiplication of teak by tissue culture was first achieved by the scientists of NCL, Pune around twenty years back. Multiple shoot formation was induced from excised seedling explants and also from excised terminal buds of 100 year old elite teak tree on a defined medium containing 6-benzylaminopurine and kinetin. Individual shoots obtained on this medium were excised and induced to root on a low salt medium containing tree auxins. These plantlets were then transferred to pots in a greenhouse and later to the field. Their growth and root system with a tap root and laterals were similar to those of plants raised from seeds. Explants from the nodal region of these sterile rooted plants were also subcultured to fresh medium where they proliferated to form multiple shoots which could again be rooted by the same procedure. Regeneration by this method has been carried out for 16 subcultures with seedlings and six subcultures with 100 year old elite tree. 50 plants have so far been obtained from the latter and are being grown in a glasshouse. It was estimated that by subculture 500 viable plants can be obtained from a single bud of a mature plant or 3000 plants from a seedling in a year (Gupta *et al*, 1980).

References

References

1. Akindele, S. O. (1989). Teak yields in the dry lowland rain forest area of Nigeria. Jour. Trop. For. Sci. 2 (1) : 32 - 36.

2. Alexopoulos, C. J. and C. W. Mims (1983). Introductory Mycology. Wiley Eastern Ltd. New Delhi, Calcutta, Bangalore, Bombay.

3. Anderson, F. (1970). Ecological studies in a Scanian woodland and meadow area. Southern Sweden. II. Plant biomass, primary production and turn over of organic matter. Botaniska Notiser. 123 : 8 - 51.

4. Anon. (1997). Nursery establishment study. Fortech Report, UPFP, U. P. Forest Department.

5. Bapat, A. R. and M. M. Phulari (1995). Teak fruit treatment machine - A prototype - II. Indian For. 121(6):545-549.

6. Bhat, K. M. (1995). A note on heartwood proportion and wood density of 8 - year old teak. Indian For. 121(6):514-517

7. Centeno, J. C. (1996). World record on teak yield : Truth or Trickery ? jcenteno@cieno.ula.ve.

8. Centeno, J.C. (1997). The management of teak plantations. http://www.ciens.ula.ve/~jcenteno.

9. Chaturvedi, A. N. (1995). The viability of commercial teak plantation projects. Indian For. 121(6):550-552.

10. Chaubey, O. P.; G. P. Mishra and Ram Prasad (1988). Phytosociological studies of teak plantations and mixed natural forests in Madhya Pradesh. II. Distribution, species diversity, productivity and some quantitative parameters of ground flora. Jour. Trop. For. 4(2):177-187.

11. Dagar, J. C.; G. Singh; N. T. Singh and G. Singh (1995). Evaluation of crops in agroforestry with teak (*Tectona grandis*), maharukh (*Ailanthus exelsa*) and tamarind (*Tamarindus indica*) on reclaimed salt affected soils. Jour. Trop. For. Sci. 7(4):623-634.

12. Dhanesh Kumar, P.; N. Rajesh; A.V. Santosh Kumar; K. Vidyasagar and M.A. Anaz (1997). Crown diameter/Bole diameter relationship as an aid to thinning in teak (*Tectona grandis* L.). Indian Jour. For. 20(4): 355-361.

13. Durga, V. V. K. and Sanjay Gupta (1995). Effect of vesicular arbuscular mycorrhizae on the growth and mineral nutrition of teak (*Tectona grandis* Linn. F.). Indian For. 121(6):518-527.

14. Egunjobi, J. K. (1974). Litter fall and mineralization in *Tectona grandis* stand. Oikos. 25:222-226.

15. Faruqui, O. (1972). Organic and mineral structure and productivity of plantation of Sal (*Shorea robusta*) and Teak (*Tectona grandis*). Ph. D. Thesis. BHU, Varanasi.

16. Fernando, S.N.U. (1966). Fertilization of teak nurseries. Ceylon For. 7(3/4):103-106.

17. Gangopadhyay, S. K.; S. Nath and S. K. Banerjee (1987). Nature and properties of some introduced teak (*Tectona grandis*) growing soils of North-West Bengal. Indian For. 113(1):65-72.

18. George, M. and G. Verghese (1992). Nutrient cycling in *Tectona grandis* plantation. Jour. Trop. For. 8 (2) : 127 - 133.

19. Gogate, M. G.; U. M. Farooqui and V. S. Joshi (1995). Sewage water as potential for the tree growth - A study on teak (*Tectona grandis*) plantation. Indian For. 121(6):472-481.

20. Gupta, D. K.; A. L. Nadgir; A. F. Mascarenhas and V. Jgannathan (1980). Tissue culture of forest trees : Clonal multiplication of *Tectona grandis* L. (teak) by tissue culture. Plant Sci. Lett. 17:259-268.

21. Hase, H. and H. Foelster (1983). Impact of plantation forestry with teak (*Tectona grandis*) on the nutrient status of young alluvial soils of West Venezuela. For. Ecol. Manage. 6(1) : 33-57.

22. Jha, K. K. (1995). Structure and functioning of an age series of teak (*Tectona grandis* Linn.) plantations in Kumaun Himalayan Tarai. Ph. D. Thesis. Kumaun University, Nainital, India.

23. Jha, K. K. and Chandra Gupta (1991). Intercropping medicinal plants with poplar and their phenology. Indian For. 121(7):535-544.

24. Jha, K. K. and J. S.Singh (1997). Temporal pattern of growth in bole volume and biomass of young teak plantation. Jour. Sust. For. (communicated).

25. Kadambi, K. (1993). Silviculture and management of teak. Natraj Publishers. Dehradun.

26. Karmacharya, S. B. and K. P. Singh (1992). Biomass and net productivity of teak plantation in dry tropical region of India. For. Ecol. Manage. 55 (1-4) : 233 - 247.

27. Kandaswamy, D.; M. Rangarajan; G. Oblisamy and R. Narayanan (1988). Mycorrhizal occurence and their effect on certain forest plant species. In : Mycorrhizae for green Asia. Proc. First Asian Conf. Mycorrhizae. Eds A. Mahadevan, N. Raman and K. Natrajan. CAS, Botany, Univ. Madras, Madras.

28. Kaul, O. N. ; D. C. Sharma ; V. N. Sharma and B. P. L. Srivastava (1979). Organic matter and plant nutrients in a teak (*Tectona grandis*) plantation. Indian For. 105 (8) : 573-582.

29. Keogh, R. M. (1979). Does Teak have future in tropical America. Unasylva. 31 (126) : 13-25.

30. Keogh, R. M. (1987). The Care and Management of Teak (*Tectona grandis* L. F.) Plantation. A practical field guide to foresters in The Carribean Central America, Venezuela and Colombia. ISBN 0 9508720 1 6. GORTA, Ireland.

31. Khullar, P. (1995). Editorial. Indian Forester. 121(6):445.

32. Khullar, P.; R. C. Thapaliyal; B. S. Beniwal; R. K. Vakshashya and A. Sharma (1991). Forest Seed. ICFRE, New Forest, Dehradun.

33. Kinhal, G. S. (1995). Technical and financial evaluation of green equities. Indian For. 121(6):566-572.

34. Kulkarni, V. (1994). Can money grow on trees ? The Economic Times on Sunday, Bangalore Sunday. 1 May 1994.

35. Lahiri, A.K. (1974). Preliminary study on rooting of green wood cutting of teak. Indian For. 100(9): 559-560.

36. Lahiri, A.K. (1989). *Taungya* based agroforestry trials in West Bengal. Indian For. 115(3): 127-132.

37. Larson, B.C. and M.N. Zaman (1985). Thinning Guidelines for Teak (*Tectona grandis* L.). Malaysian For. 48(4): 288-297.

38. Macdonald, B. (1986). Practical woody plant propagation for nursery growers. Timber Press. Portland, Oregon.

39. Negi, J. D. S. ; V. K. Bahuguna and D. C. Sharma (1990). Biomass production and distribution of nutrients in 20 years old teak (Tectona grandis) and gamhar (Gmelina arborea) plantation in Tripura. Indian For. 116(9) : 681-686.

40. Negi, M. S. ; V. N. Tandon and H. S. Rawat (1995). Biomass and nutrient distribution in young teak (Tectona grandis Linn. F.) plantations in Tarai region of Uttar Pradesh. Indian For. 121 (6): 455 - 464.

41. Negi, S. S. (1996). Teak (Tectona grandis). Bishen Singh Mahendra Pal Singh, Dehradun.

42. Ola - Adams, B. A. (1993). Effects of spacing on biomass distribution and nutrient content of Tectona grandis Linn. F. (teak) and Terminalia superba Engle. & Diels. (agara) in south - western Nigeria. For. Ecol. Manage. 58(3-4):299-319.

43. Parameswarappa, S. (1995). Teak - How fast can it grow and how much can it pay ? Indian For. 121(6):563-565.

44. Prasad, R. S. R. (1995). Tectona grandis - Elite management. Indian For. 121(6):558-562.

45. Puri, S. and K. S. Bangarwa (1990). Mass culturing of trees by vegetative propagation technique for afforestation of wastelands. Proc. Nat. Sem. on Technology for Afforestation of Wastelands. FRI, Dehradun.

46. Purkayastha, S. K. (1985). Indian woods. Vol. V. Manager, Govt of India Press, Nasik.

47. Raman, N. and S. Gopinathans (1992). Association and activity of vesicular arbuscular mycorrhizae of tropical trees in tropical forest of Southern India. Jour. Trop. For. 8:311-322.

48. Rawat, J. K. (1995). Value of a 20 year old irrigated teak plantation. Indian For. 121(6):553-557.

49. Reddy, C. J. (1995). The bounty from teak tree. Indian For. 121(6):573-575.

50. Sarlin, P. (1966). The first thinning of Teak plantation. Bois. For. Trop. 108:5-20.

51. Satoo, T. (1970). A synthesis of studies by the harvest method : Primary production relations in the temperate deciduous forests of Japan. In : Analysis of temperate forest ecosystem. pp 55-72. Ed. D. E. Reichle. Springer Verlag, New York.

52. Seth, S. K. and O. N. Kaul (1978). Tropical forest ecosystem of India : The teak forest. In : Tropical forest ecosystem. UNESCO, Paris.

53. Sharma, A. and M. L. Naik (1989). Biomass and productivity studies in teak (*Tectona grandis* Linn. F.) under artificial plantation in Surguja district (M.P.). Trop. Jour. For. 5(3) : 97-104.

54. Sinha, B. K. P. (1996). State Forestry Action Programme. Uttar Pradesh Forest Department.

55. Singh, A. K. and B. N. Gupta (1993). Biomass production and nutrient distribution in some important tree species on Bhatta soil of Raipur (Madhya Pradesh) India. Ann. For. 1(1) :47-53.

56. Singh, Lalji and J. S. Singh (1991). Storage and flux of nutrients in dry tropical forest in India. Ann. Bot. 68 : 275-284.

57. Singh, K. P. and R. Mishra (1979). Structure and functioning of natural, modified and silvicultural ecosystems in Eastern Uttar Pradesh. 160 pp. Final technical report (1975-1978). MAB research Project, BHU Varanasi.

58. Singh, S. B.; S. Nath; D. K. Pal and S. K. Banerjee (1985). Changes in soil properties under different plantations of Darjeeling forest division. Indian For. 111(2):90-98.

59. Srivastava, S. S. (1997). Teak Management Under Commercial Forestry. Personal communication.

60. Stern, K. and L. Roche (1974). Genetics of forest Ecosystem. Springer Verlag. Berlin, Heidelberg, New York.

61. Tandon, V. N.; M. S. Negi; D. C. Sharma and H. S. Rawat (1996). Biomass production and nutrient cycling in plantation ecosystem of *Eucalyptus* hybrid in Haryana. 2. Distribution and cycling of nutrients. Indian For. 122(1):30-38.

62. Tewari, D. N. (1992). A Monograph on Teak (*Tectona grandis* Linn. F.). I. B. H., Dehradun, India.

63. Uniyal, D. P. and M. S. Rawat (1995). Effect of temperature and relative humidity on grafting and budding of teak (*Tectona grandis* Linn. f.). Indian For. 121(6):510-513.

64. Verma, R. K. and Jamaluddin (1995). Association of arbuscular mycorrhizae of teak (*Tectona grandis*) in Central India. Indian For. 121(6): 533-539.

65. Weaver, P. L. (1993). *Tectona grandis* L. F. SO - ITF - SM - 64 . September 1993.

66. Weaver, P. L. and J. K. Francis (1992). Performance of *Tectona grandis* in Puerto Rico. Commonwealth For. Review. 69(4):313-323.

Subject Index

Subject Index

Subject Index